IN YOUR GARDEN

TELETHON

JEFF DORRINGTON was born at an early age in 1950 and in spite of himself gained a sort of education, none of which has proved useful in later life. He naively opened a bookshop in Midland in 1971, with his mother as a partner. A total lack of business knowledge or bookselling skills assured its success, and it is still a thriving bookshop today.

Jeff acquired his love of gardening from his father, who could smell a horse turd at fifty metres and smuggle it home to the garden. In 1990, with incredible gall, no business plan and his innocent wife Anne-Marie as a partner, Jeff started The Rare Tree Company. His desire to grow the many beautiful, and sometimes difficult to obtain, plants this world has to offer has led to his well-proven experience of their needs in our mild climate. The business has blossomed to encompass two nurseries occupying over six acres of production and resale areas.

Jeff's very successful gardening series, incorporated in Channel 7's 'Today Tonight' program, led to this more tangible form of what can be grown from the coast to the hills of Perth.

IN YOUR GARDEN
WITH
JEFF DORRINGTON

FREMANTLE ARTS CENTRE PRESS

First published 1997 by
FREMANTLE ARTS CENTRE PRESS
193 South Terrace (PO Box 320), South Fremantle
Western Australia 6162.

Consultant Editor Alex George.
Designer John Douglass.
Production Coordinator Cate Sutherland.

Typeset by Fremantle Arts Centre Press
and printed by Sands Print Group.

National Library of Australia
Cataloguing-in-publication data

Dorrington, Jeffrey Lee, 1950- .
      In your garden.

      ISBN 1 86368 202 3

      1. Gardening - Western Australia - Perth Metropolitan Area -
      Handbooks, manuals, etc. I. Title.

635.099411

A
artswa

The State of Western Australia has made an investment
in this project through the Department for the Arts.

CONTENTS

Introduction   7

Gardening in Perth   11

Trees   16

Shrubs   74

Creepers and Climbers   146

Photographs   156

References   156

Index   157

ACKNOWLEDGEMENTS

My grateful acknowledgement to my mother who always believed in me; my wife, Anne-Marie, who has always stood beside me; Alex George — a kind and diligent editor; and Ivan Zar and Dave Hole, whose music helped me through the long nights.

Cover photograph by Frances Andrijich.

Internal photographs by Don Ellison, except as follows: John (Greenfingers) Colwill, pp. 30, 65, 69, 71, 71, 74-75, 82, 84, 95, 96, 111, 141, 146-147; John Douglass, pp.16-17, 77; Jeff Dorrington, pp. 33, 46, 54, 62.

Sketches by Shane Tholen.

INTRODUCTION

Gardening was once described, by a long-term English gardener, as eleven months of hard work and one month of bitter disappointment. In recent times, however, gardening has changed so dramatically that we no longer need suffer the heartache of our English friend.

Garden technology is an almost unsurpassed growth industry. Our biggest problem is staying abreast of new inventions, new plant introductions or improvements to existing systems. The Internet will be the best, the most accurate and current directory of this information – interestingly this electronic library won't require the pulping of trees to spread information.

This book has been compiled with the non-professional gardener in mind – someone who loves gardening, knows what he or she wants to achieve but may not know which tree or shrub would fit the bill. The aim of the plant list is to lead the gardener alphabetically through a range of delightful possibilities, in most cases by common name, except for some species which are listed by their better-known botanical name. Most of us know plants only by their common names, but with each entry you will find the correct botanical name in round brackets, and, in a few cases, an alternative botanical name following that in square brackets. When seeking a plant for your garden, it is best to request it by its botanical name. The Latin (or Latinised Greek) name used is universal and unique to that plant, thereby removing the chance of a mix-up – provided that the nursery has the correct name! The names are those in current use in Australia.

Estimating sizes of trees has always presented a problem, as some don't stop growing. Therefore, I have estimated all sizes at what could be expected from a tree ten years old. Many books estimate a tree's size either in the country of origin or where the author has experienced it growing. In our case these measurements relate directly to the growing conditions in the south-west of Western Australia.

As a keen gardener for many years, I have found that there is a wealth of information about native trees and shrubs but very little about the exotics and how they perform in Perth and the general south-west region of Western Australia. This situation has changed very little, and so the main thrust of this book is to redress that problem. Some exotics such as abelia and coprosma have been grown in Western Australia for so long that many may think they are native. These, too, have been left out in favour of lesser known but equally hardy plants.

'Hardiness' in horticultural terms means a plant's ability to withstand cold in various geographical zones. In Western Australia, however, where cold is rarely a problem, we refer to different forms of hardiness. Our problems are the ability to withstand summer heat and to require as little water as possible but still perform well. Necessity being the mother of invention has produced some wonderful ways of combatting our heat and water problems; these will be dealt with in the next section — 'Gardening in Perth'.

Too few people realise that their residence is their largest investment and best asset — and 50% of that asset is the garden. Those homes with well kept or well presented gardens always sell first. The very first impression a purchaser will get is the garden. Although new carpets may look fabulous, the sale may have been already lost because of the garden. I always find it interesting that some people will spend thousands of dollars on new carpets and curtains and yet throw their hands up in horror when a few hundred dollars' worth is recommended for the garden. Sure, the plants may die if not looked after — but have you ever tried getting a spilled cup of strong black coffee out of white shag pile? Life is made up of risks.

The Nursery Industry Association of Western Australia (NIAWA) is the strongest it has been in many years. The Association administers a voluntary code of nursery hygiene which is designed not only to reduce the incidence of dieback (*Phytophthora cinnamomi*) but also to ensure that disease-free plants are made available to a discerning purchasing public. Members of the Association are graded in a manner similar to the way the hotel industry uses stars to indicate the standard thus far achieved. NIAWA members are easy to spot by an identifying sign, and the quality of greenstock will be obvious. Nurseries displaying this sign need your support as this encouragement can only help achieve the high quality for which we all are striving.

The open garden scheme has been operating in Western Australia for a while now. This is such a good way to benefit from the experiences of local gardeners and have a stickybeak at the same time. I have seen many of the gardens now and am amazed at the wealth of ideas, innovations, styles and range of gorgeous plants, all in Western Australia. We have a gardening heritage we can be proud of and a bunch of dedicated gardeners leading us into the next century.

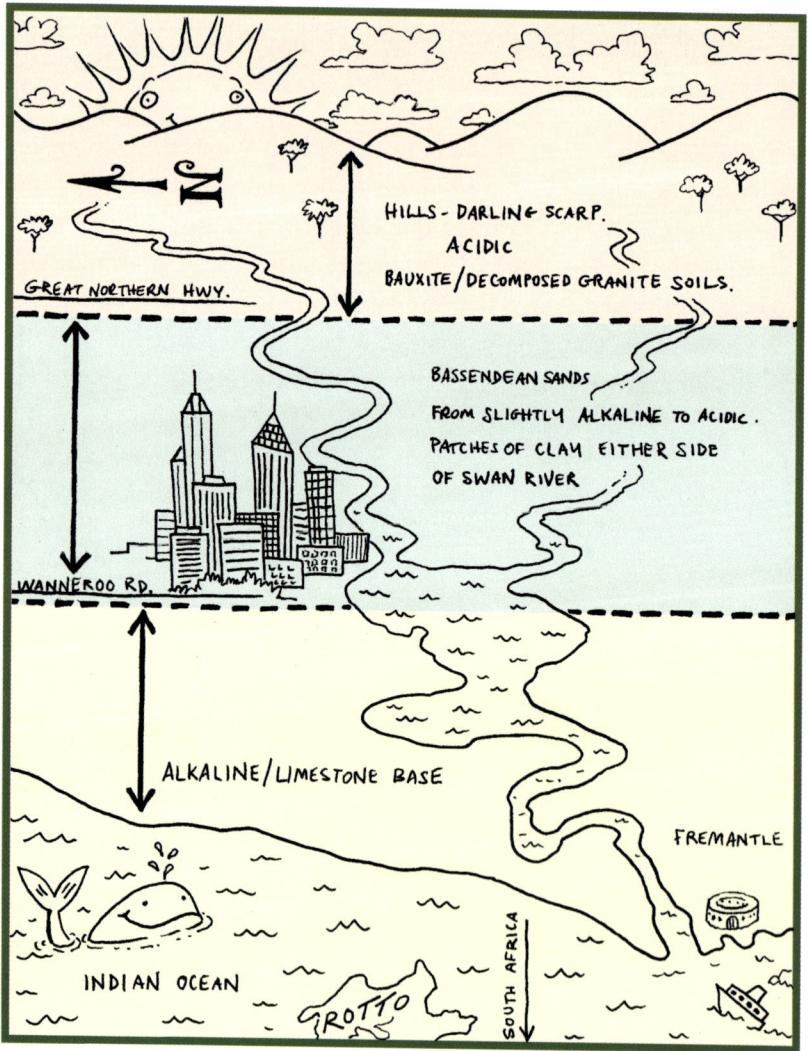

HILLS - DARLING SCARP.
ACIDIC
BAUXITE/DECOMPOSED GRANITE SOILS.

GREAT NORTHERN HWY.

BASSENDEAN SANDS
FROM SLIGHTLY ALKALINE TO ACIDIC.
PATCHES OF CLAY EITHER SIDE
OF SWAN RIVER

WANNEROO RD.

ALKALINE/LIMESTONE BASE

FREMANTLE

INDIAN OCEAN

ROTTO

SOUTH AFRICA

GARDENING IN PERTH

The climate is Mediterranean — hot summers, cold winters, good if not plentiful water — and much of our soil is dreadful! Bearing all this in mind, we need to do two things: first, choose plants that will do well in the climate, and second, try to remedy the soil. If you get it right with the soil you will at the same time conserve water.

The big secret is putting humus into the soil. Humus is organic matter that, over time, breaks down, encouraging necessary micro-organisms which in turn release nutrients to your plants — and — helps in the retention of water. I don't throw away anything! In the kitchen we have a chook bucket and all scraps go to the chooks; later I collect the fowl manure and use it in the garden. If you are not lucky enough to have a group of contented chooks, you certainly can have a compost heap or bin and return all those scraps to your garden. Any thriving garden will have large amounts of humus added to it regularly; once is not enough — it must be regular. You will also notice that worms take up residence in this humus and they will do all the work for you, taking the goodies down to the root zone of your plants.

My most successful plants are those that I mulch deeply. After many years of experimenting, I have come to the conclusion that you can use anything as long as it is organic. Here is my method. First a thick wad of newspaper (after you've read the comics), then a layer of sheep or cow poo (chook is fine if it is well rotted and you don't live on limestone) about 10 cm deep and then around 10 cm of the chewed up leaves and branches the big roadside mulchers make. Water it well and you will be amazed at how the soil retains that moisture under the newspaper.

The purpose of the poo is not only to add nutrients but, more importantly, to help break down the newspaper and mulch. As they slowly start to compost, nitrogen from the soil will be withdrawn to be used in the composting process. As nitrogen is vital to your plants' healthy growth, they will miss out until the composting is completed, at which point it is returned to the soil. So if you use a high-nitrogen source in your deep mulching — poo — those micro-organisms will have it right on hand and your plants won't miss out.

When it comes to mulch you can use any organic matter. Cotton, wool, carpet (not synthetic), paper, lawn clippings, leaves, and of course the goodies from your compost heap. As you apply the mulch the nature of your soil will change. Even if you live on the famous Bassendean Sand or up in the hills with gravelly clay, you can create a lovely rich, moisture-retentive soil.

People living on or close to the coast will find that their soil type is alkaline due to the amount of limestone. This makes it difficult to grow acid-loving plants such as camellias, azaleas and the like. The use of acidifying agents will help convert the soil, and cow manure is especially useful in mulching as it tends to be more acidic. Chook poo is strongly alkaline and will make the situation worse, thus it must be used with care. The reverse is true of the hills where we have a naturally acidic soil, but as most plants favour the acidic side many things will grow happily in this soil. Chook poo is useful if you want to plant alkaline-loving plants or grow pink hydrangeas. That middle slab of Bassendean Sand ranges in its pH from mildly alkaline to acidic — so it would be worth doing a quick pH test to find which way yours is. Inexpensive pH test kits are available from most nurseries.

Spend any time in the nursery industry in Western Australia, and when it comes to trees three questions are constantly asked: how high? how wide? what are the roots like? I can't help thinking that someone many years ago frightened Perth gardeners so much that these questions are repeated as some sort of litany. While it is important to secure the right-sized tree for the garden, it is worth travelling to Victoria, New South Wales or Canberra to see gorgeous, cool, leafy gardens full of trees. They don't have the fear of trees that we seem to have — in fact it prompted the famous Australian architect Robin Boyd to say that Western Australians suffer from 'arborophobia' — a fear of trees. Although we have some magnificent tree specimens and some very leafy suburbs around Perth, we have nowhere near enough trees! The ability of trees to reduce the heat in summer is monumental. Children can play under and in them without the fear of sunburn, dogs and cats always take advantage of their shade, we can sit out under them in the cool of the evening, and you can cut massive evaporation from your garden by planting more trees. The cooler your garden, the more you will want to spend time in it.

Added to the fear of trees is the fear of mess! All trees shed their spent leaves. Leaves lying on the ground seem to bring out the worst in some people, but I rather like the look of the naturalness of leaves falling — perhaps I'm a bit messy! If you are one of those who do not like leaves falling throughout the year, then I recommend that you consider some deciduous trees. These are the trees that shed all their leaves in late autumn — many changing to beautiful colours of the red part of the spectrum first — the leaf drop usually takes two weeks and that's it for the year. Definitely the 'cleanest of trees'. Many of this group have spectacular flowers as well, a bonus.

The evergreens are a group that includes most conifers and most native trees, although there are many from other countries as well. These trees shed their leaves throughout the year and could be considered the 'messiest' trees. The evergreens do have distinct advantages, those of screening unsightly views and giving privacy all year round. Many of these trees also produce brilliant shows of blossom, although rarely autumnal colour.

Whatever type of tree you plant, it is essential that you water it correctly to encourage deep root growth and remove any possibility of invasive roots. All trees should be hand watered; do not rely only on reticulation. Most forms of reticulation are used for shallow-rooted plants such as lawn or shrubs and, unless left running for a long time, will water to a very shallow depth. If the roots on your tree can't find the water they need deep down, they may turn around and come back up to where the water is — and that's when the problems start. So when you plant a tree or large shrub, make a well around the base 40 cm from the trunk and hand water for at least two summers. In mid summer this should be about three times a week with a long soak. I measure this by the duration of drinking a can of beer per tree, but I am sure a gin-sling would do!

Whatever you add to your garden will require fertilising. It is important to know that manure is not a fertiliser, but rather a nutrient-based form of humus. Fertiliser is a balanced blend of all the nutrients that a plant needs to maintain healthy growth. All fertilisers will have, in different proportions, the three main elements: nitrogen, phosphorous and potassium, known as N:P:K (their chemical symbols). Following that, if you purchase an all-purpose fertiliser, will be the trace elements. N:P:K will be by far the greater proportion in this mix, some of the trace elements being in minute quantities. The all-purpose fertilisers have been developed to save you the trouble and expense of formulating your own, and they work extremely well. If you are growing Australian natives you need to be wary of adding too much phosphorous as our soils mostly lack this element and the natives have adapted to cope with this. There are special 'native' formulations on the market that cater for this group. Slow-release fertilisers have a coating that disintegrates at different rates to allow their nutrient release to span a certain length of time. I find an application of eight–nine month slow-release will last from spring through to the winter dormancy of most plants, and there are several brands on the market to choose from. Both all-purpose and slow-release fertilisers are a balanced blend of major and micro-elements needed by your plants. Specialist fertilisers are also available, for instance rose, citrus or camellia mixes. These are excellent for their respective groups and should be used only for the specific application.

Potting mixes are many and varied. I thoroughly recommend a brand that carries the Australian standards symbol. That way you can be assured of the rigorous testing the mix must go through to achieve this mark. Cheap potting mixes are hardly worth the bag they are put into — a little like putting a fifty dollar plant into a one dollar hole; when you've gone to that much expense why risk losing it for a few lousy dollars? What you need to look for in a good potting mix is drainage, its ability to last the distance, the nutrients that have been added, and an additive such as coarse pine bark to make sure it doesn't collapse and compact in a couple of months.

To close this section I would like to note the service nurseries and garden centres offer the experienced and novice gardener. Not only do they go to considerable trouble and expense to locate the best and healthiest plants to make available to us, but also they train staff to offer the best advice, where to plant, when to plant and how to remedy a problem. You will find an enormous range of associated items constantly in stock to satisfy the most demanding gardener. Some nurseries specialise, such as in bonsai or camellias — but whatever the size or area of specialisation you will find someone who not only knows his onions but loves them. That must count for a lot. If we lose them to businesses dedicated to making money as their prime motive we may well regret it.

TREES

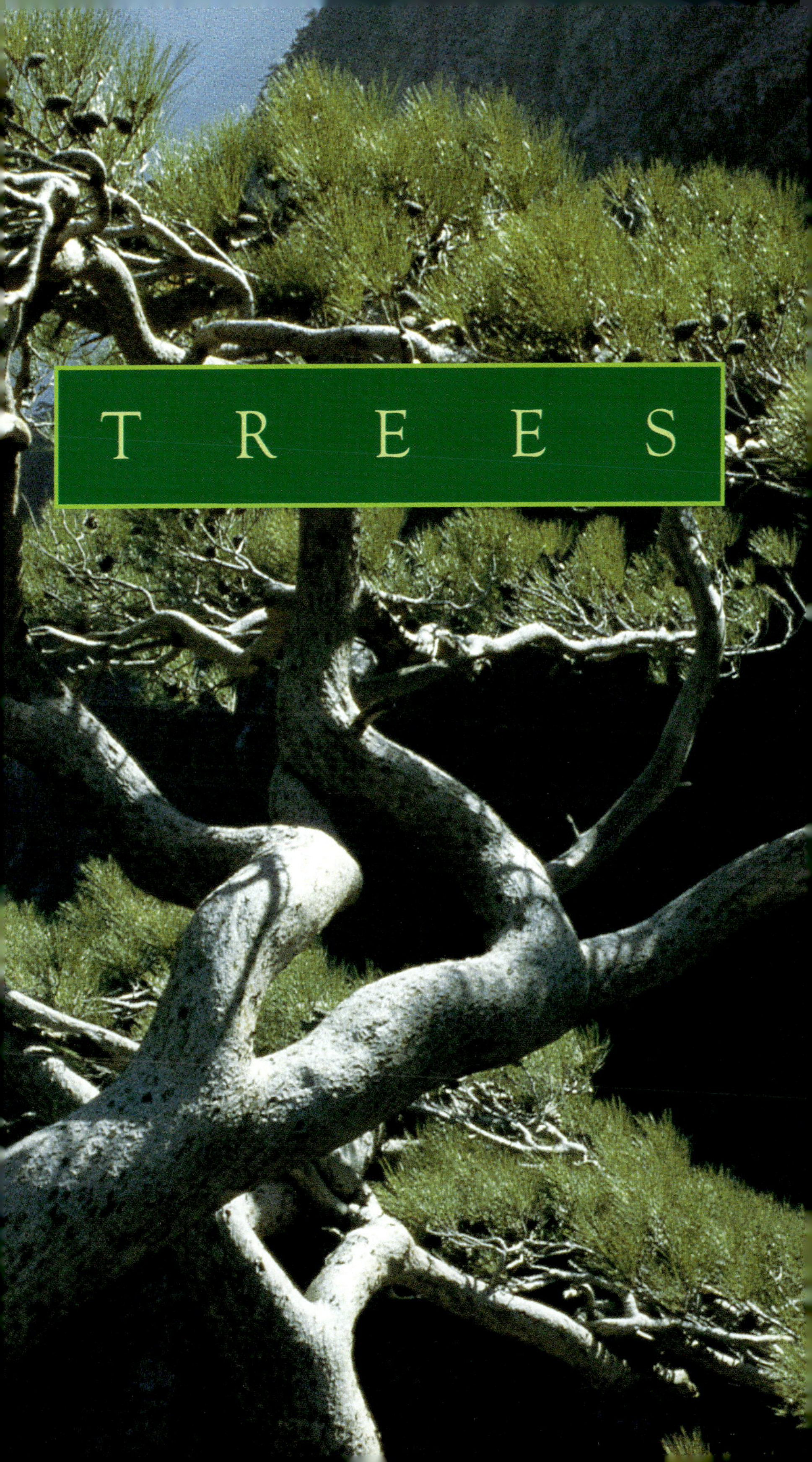

ALBIZZIA

(*Albizia julibrissin*) also known as Persian Silk Tree

FEATURES: Deciduous; height to 4–5 m; spread to 4–5 m giving dappled shade. Non-invasive root system.

FLOWERS: Soft to intense pink, like an upturned shaving brush about the size of a twenty cent coin. Profuse flowering over summer. Produces seed pods.

LEAVES: Small, bipinnate with paired leaflets. Very attractive, turning coppery-bronze in autumn.

WHERE: Can be grown anywhere on the coastal plain. This tree appears to be tolerant of most conditions. Will grow in the hills but may need protection from frost when young.

USES: Mainly as an ornamental, but is useful for providing shade for sub-tropical species that need an upper canopy cover.

COMMENTS: A very beautiful small tree to introduce lush foliage and a rich pink over summer. Best bottom-pruned to remove lower branches and encourage a single straight trunk to 2 m. Its ability to acclimatise almost anywhere and the ease of growing means it should be grown in many more gardens.

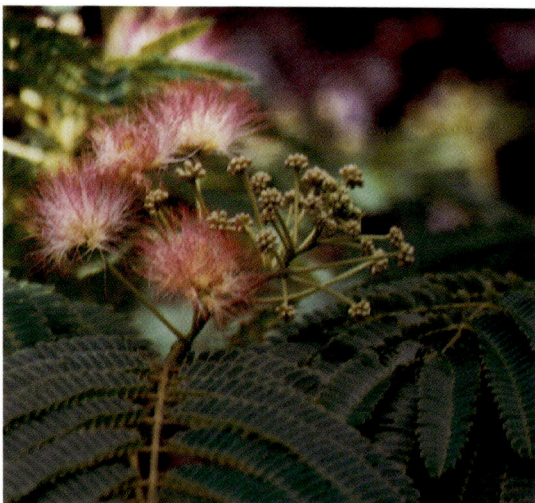

HANDY HINT

CAUTION. If you must buy plants in a supermarket, check to make sure that they are not stored in refrigerated airconditioning or indeed offered for sale there. This form of airconditioning, whilst nice for us, is devastating to plants. It sucks all the moisture out of them, and the longer they are in there the worse it is. Quite commonly, they get an enormous shock when taken outdoors and may not survive.

BAUHINIA

(*Bauhinia variegata* 'Candida') also known as White Bauhinia [*B. alba*]

FEATURES: Deciduous; height to 5–6 m; spread to 4–5 m. Non-invasive.

FLOWERS: Described as the 'poor man's orchid'. Beautiful white flowers mass over the tree before new leaf set. Flowering time is variable but they usually flower in very late winter through to late spring.

LEAVES: Butterfly-shaped. Can be quite large and distinctive. Named after twin botanists called Bauhin because the leaf is deeply lobed giving a matched pair of leaflets.

WHERE: Very adaptable to all conditions from the coast to the hills. Loves a hot spot in the garden, but water well over summer.

USES: Mainly as an ornamental. Will tolerate strong wind although there could be some leaf burn if not watered well. Makes a good screen; there are some spectacular standards on the river side of Wesley College in South Perth.

COMMENTS: The flowers look great at a distance, being a mass of white, but are also remarkably beautiful up close. Easily grown and was used extensively by early gardeners.

VARIETIES: Hong Kong Orchid (*Bauhinia x blakeana*). White and mauve colours in the same flower, otherwise the same.

Bauhinia galpinni is a low shrub with rich orange flowers over early summer. Makes an excellent hedge. Good examples can be seen around the administration block of Ascot race track.

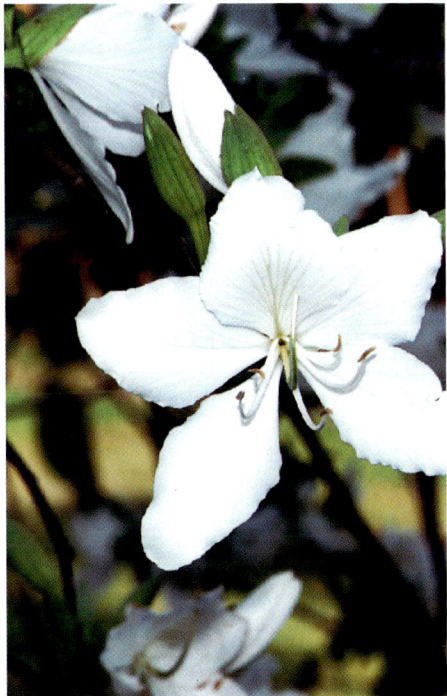

BOX ELDER MAPLE

(Acer negundo)

FEATURES: Deciduous; height to 7 m; spread to 4 m. Non-invasive.

FLOWERS: Very unusual. They look like a small bunch of whitish-pink filaments or threads that hang downwards. Then the winged seeds are formed, but these are commonly sterile as this tree is dioecious and is often not pollinated as only one sex is grown.

LEAVES: Elongated, oval. Not densely leafy but gives good dappled shade.

WHERE: Away from fierce winds. From the coast to the hills if protected from westerlies or easterlies. Likes a rich, moist, cool root run, which with maturity it will provide for itself, but deep mulch when young.

USES: A famous tree for its shape and overall appearance. The tree has a rounded canopy and a flat bottom — just like the ice cream on top of a cone. Mostly ornamental, although it is used for timber in Japan.

COMMENTS: Just outside the Butterfly House at the Perth Zoo is a typical example of this attractive tree. Often the cultivars are grown instead because of the variegation and or leaf colour

CULTIVARS: 'Golden Variegated Box' Maple has gold and green leaves. A very pretty small tree when it reaches a bit of height.

'Silver Variegated Box' Maple has white and green leaves. When a gentle wind moves them they shimmer with a silvery sheen.

'Kelly's Gold' is a solid gold leaf from spring to autumn. This particular cultivar has a pride of place in our garden — I just love it.

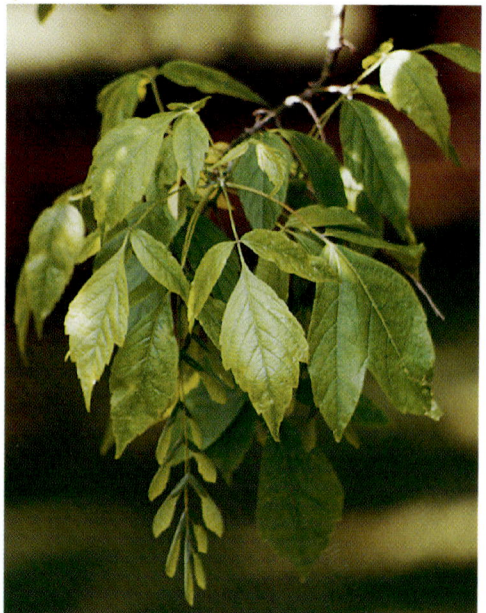

BULL BAY MAGNOLIA

(*Magnolia grandiflora* 'Exmouth')

FEATURES: Evergreen; height to 10 m; spread to 5 m. Non-invasive.

FLOWERS: Huge and heavily perfumed — as big as a cup and saucer. An old-fashioned tree that has never fallen out of favour, and quite rightly. The flowers are stunning floated on water in a bowl as a centrepiece on a table. Flowers appear in early summer.

LEAVES: Large and deep green. A good foil for the flowers, making them stand out. Will cast dense shade.

WHERE: Coastal suburbs through to the hills. There is a very mature, handsome specimen in Queens Gardens in East Perth.

USES: In the USA they are used as street trees — what a treat for all the people living there. They are fine under power lines as their relatively slow growth means that they will rarely require pruning.

COMMENTS: The cultivar 'Exmouth' has a pronounced brown furry underside to the leaf and flowers young. By far the best one to grow — but check under the leaves first! Anyone who has smelled the perfume of this tree will want one for themselves. It is heavenly.

CHINESE PISTACHIO

(Pistacia chinensis)

FEATURES: Deciduous; height to 4–5 m; spread to 2 m. Non-invasive.

FLOWERS: Insignificant.

LEAVES: Dark green, paired; looks cool and lush over summer. The autumn colours are scarlet and crimson, very deep, almost translucent.

WHERE: A very tough tree that copes with all sorts of conditions. Likes a good soil, well prepared with humus. Shows drought-tolerance but responds well to a good long soak. Takes the easterlies well.

USES: A small tree that fits well into cottage gardens, or can be used to create a cool lush look in the garden. Excellent for courtyards. Looks lovely if trained up and bottom pruned to around 2 m; the shape then becomes quite rounded.

COMMENTS: An excellent substitute for the Rhus tree which has similar autumnal colour but can produce a rash on susceptible people and therefore is rarely grown. The Chinese Pistachio has none of these problems. This is not the fruiting variety that we all love to eat but an ornamental form. In autumn, it can be breathtaking in its beauty, and interestingly is one of the last trees to colour — therefore can be used to extend autumn colour in the garden.

HANDY HINT

TERRACOTTA POTS. Terracotta has one major drawback when used for pots — it is very porous and absorbs the water intended for the plant, especially if it has been allowed to dry out. To fix this problem you can paint the inside with a sealing agent or melted candle wax or simply line the pot with newspaper. If you paint the inside, cover only two-thirds of the inside area so that drainage may still occur.

CHINESE TALLOW

(Sapium sebiferum)

FEATURES: Deciduous; height to 6–7 m; spread 3–4 m. Non invasive.

FLOWERS: Small, yellow, standing up like little candles, but nothing very special.

LEAVES: Heart-shaped, hanging downwards. Very attractive over spring and summer, but in autumn this tree puts on a spectacular show of colour. You can expect colours from yellow to deep crimson and every shade between. The tree fairly glows and will carpet a lawn with the same rich colours.

WHERE: Having a compact root system, the Chinese Tallow can be grown in a courtyard from the coast all the way through to the hills. In the garden they give good shade, are wind-tolerant and make a nice screen. Soil types don't seem to worry this tree, although a good mulching and watering will bring out its best.

USES: Good in tight areas where an overhead canopy is required for shade, for example a courtyard. The tree branches naturally from the ground up and needs bottom pruning if it is to develop a trunk — not hard to do.

COMMENTS: I often feel that if Leonardo da Vinci had designed a tree for Perth gardens, this would be it. Hard to find a fault with it. One of my favourites — especially in autumn.

CLARET ASH

(Fraxinus oxycarpa 'Raywood')

FEATURES: Deciduous; height to 10-12 m; spread to 5-6 m. If watered deeply and regularly over the first two summers, roots will develop deeply.

FLOWERS: Nothing special, often difficult to see.

LEAVES: Deep green from spring through summer. In autumn, the leaves slowly start to develop a deep red hue. This colour accelerates until the entire canopy is a deep claret red. The leaves offer very good summer shade.

WHERE: The Claret Ash can be found growing virtually on the beach front at Mandurah, through all suburbs and up into the hills.

USES: Commonly used in street plantings.

COMMENTS: Fast-growing, very attractive tree. The height can be reduced by removing the main leader to encourage branching, and plants are commonly sold this way. Tolerant of any conditions. The wine-red leaves in autumn give colour contrast in any garden, and summer shade is unsurpassed. Fertilise in spring with an all-purpose blend.

CULTIVARS: 'Golden Ash' is a smaller, bushier tree. Its leaves are golden in spring, light green over summer and brilliant gold in autumn — makes an excellent foil when planted near a Claret Ash.

'Variegated Ash' has a soft white variegation in the green leaf and is more elongated than the other ashes. A very lovely small tree with a distinctive colour.

'Weeping Golden Ash' is similar to the Golden Ash except that it is grafted at around 2 m and forms a gorgeous circular head. A very special specimen tree.

VARIETIES: The Evergreen or Flowering Ash (*Fraxinus griffithii*) is a non-deciduous small tree. Maximum height to 4-5 m. Ideal as a screening tree to block unsightly views or neighbours' houses. It flowers at about five years with a mass of white blossom, very small flowers that cover the tree.

HANDY HINT

VENUS FLY-TRAPS. Venus Fly-traps fascinate kids (big and little) but there are some things they don't like. The first is drying out. More often than not this is the primary cause of death. Best to purchase one in a 'water-well'. This is a self-enclosed watering system with a wick at the bottom to draw water up from the reservoir. This system is nearly foolproof — just make sure the reservoir has water in it. Second, use only rainwater or distilled water on your fly-trap. The chemicals and nutrients in our drinking water are fatal to a trap. Third, keep it out of the sun; dappled light is fine. Last, they like a warm, humid atmosphere with no draught. They can feed themselves quite adequately on the odd fly that they trap, and one fly can last them for weeks. If you must feed them a fly, don't spray it first!

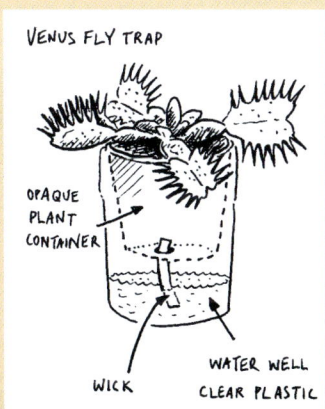

VENUS FLY TRAP

OPAQUE PLANT CONTAINER

WICK

WATER WELL CLEAR PLASTIC

DOMBEYA

(*Dombeya natalensis*) also known as Cape Wedding Bush

FEATURES: Evergreen; height to 3–4 m; spread to 2 m. Non-invasive.

FLOWERS: From late autumn through winter. Little white parachutes that hang down from the tree in bunches. Nice to have something flowering at this time. When in full flower the pure white against the very dark large green leaves looks brilliant.

LEAVES: Often of two sizes. A mix of large and smaller leaves on the perimeter of the plant that screen the interior from view. The leaves are very dark green and can cast good shade.

WHERE: They do well on the coastal plain and in the hills. Tolerant of most soil types, but will respond to good mulching and watering over summer. They are drought-tolerant, but don't look their best without a good soak.

USES: A perfect screening tree. In leaf all year round, they make a dense screen. One of the best privacy trees and, being small, rarely need pruning. Used this way the flowers are just a bonus.

COMMENTS: Originally from the Cape area of South Africa, they are well adapted to our conditions and thrive here. A very lovely bushy tree that covers from the ground up.

VARIETIES: *Dombeya macrantha* is very similar in its habit to the Cape Wedding Bush but with lighter coloured leaves and flowers over winter varying from soft pink to red.

FICUS

(Ficus benjamina)

FEATURES: Evergreen; height to 15 m; spread to 8 m. Invasive.

FLOWERS: Inconspicuous, hidden inside the figs.

LEAVES: Glossy, dark green, carried densely on a high canopy. The tightly packed leaves are a special feature.

WHERE: The fig family is renowned for its toughness, growing in the worst of situations and still looking good. They enjoy good watering, some fertilising and full sun. Having said that, they are often used as an indoor potted plant which indicates their versatility. Best grown in a container unless you have plenty of room.

USES: This tree lends itself to topiary in a big way. I have seen these trees four metres high with a huge, perfectly round head on a 2 m trunk, like a giant lollipop. Smaller versions are also very popular. The idea is to restrict the size of the roots and therefore the tree. They also make a great hedge.

VARIETIES: The Moreton Bay Fig (*Ficus macrocarpa*) is the stuff dreams are made of. I know of one in Bassendean that is now larger than the quarter acre block it was planted in many years ago. Great if you have a large property. Strangely enough, many people use them in bonsai — and they look wonderful. May be messy when the figs fall.

The Bo Tree (*Ficus religiosa*) is supposedly the tree Mohammed sat under. The leaf is very attractive, quite broad, narrowing to a long point and is used in India for painting on. This, too, will grow to a large tree unless cut back or grown in a container.

FLOWERING PEACH

(Prunus persica 'Albopena')

FEATURES: Deciduous; height to 4–5 m; spread to 2 m. Non-invasive.

FLOWERS: Double pink, dotted all over the bare stems and branches. Visible from quite a distance.

LEAVES: Typical peach, long and thin, commonly with different colours. Not much autumn colour.

WHERE: Peaches are quite hardy and look superb in a cottage garden. Suitable from the coastal plain to the hills. They benefit from an enriched soil and hand watering until established. Quite fast growing in good conditions.

USES: Great to add vibrant colour to a dull garden or to add accent to one area. By using the different cultivars and colours you can create an orchard of colour instead of fruit.

COMMENTS: The first Flowering Peach bud to burst is a definite sign of spring. It is followed closely by a mass of blossom, far more than the fruiting trees. All peaches benefit from a thorough spraying with Bordeaux mixture before flowering or leaf set, to prevent leaf curl.

CULTIVARS: 'Versicolor' has mostly white flowers with an occasional pink or red one, or just a stripe of red in one petal. Very eye-catching.

The Weeping Peach (*Prunis persica* 'Pendula') is grafted at 1.5–2 m and produces at first a horizontal head that finally weeps. Available in either crimson or 'Versicolor'.

FLOWERING PLUM

(Prunus x blireiana)

FEATURES: Deciduous; height to 4–5 m; spread to 2–3 m. Non-invasive.

FLOWERS: Double pink, larger than most plum flowers. The perfume is heavenly — reminds me of cherry.

LEAVES: Have a range of colours, depending on the heat over summer, from plum red to deep green, and sometimes a mixture of both. The autumn colour is very respectable as well. The leaf is typical plum, broad and serrated.

WHERE: Anywhere except facing the ocean. Takes all types of soils, but will perform much better if deeply mulched and well watered over summer for two years or so.

USES: Cottage gardens need this tree to help balance them. A row of them pleached is amazing. Two planted close by look wonderful and one on its own is a major statement.

COMMENTS: All around the Arc de Triomphe in Paris is a row of blireianas and the effect is just beautiful. They are so easy to grow that I think every home should have one.

SIMILAR SPECIES: Red-leaf Plum (*Prunus cerasifera* 'Nigra') is an old fashioned variety enjoying a major revival. The deep red leaf is present from spring to early winter and adds a deep colour accent to any garden.

Prunus wrightii is a New Zealand hybrid with much vigour. It produces huge pink flowers and a wonderful coloured leaf.

GLEDITSIA SUNBURST

(*Gleditsia triacanthos* 'Sunburst')

FEATURES: Deciduous; height to 8 m; spread to 5 m. Needs to be watered well over the first two summers, by hand.

FLOWERS: Small, golden, difficult to distinguish from the foliage. Insignificant.

LEAVES: Paired, small and ferny. Gold in spring, gold and green over summer, and old gold in autumn.

WHERE: A very robust tree, good from the coast to the hills. It has amazing toughness and can tolerate drought conditions. It will look much better if well watered and fertilised.

USES: Commonly used as a street tree in the Eastern States. Good under power lines. Adds accent to any garden. Not an overly large tree but its overall attractiveness makes it very worthwhile.

COMMENTS: Should be more of them. Especially effective if planted with a companion cultivar called 'Rubylace'. These trees produce seed pods, and if you are like me you will have an overwhelming desire to plant the seeds. Don't. What you will get is the rootstock, and that's where the second part of the name comes in. 'Triacanthos' means three thorns, or barbs — and they are wicked. You can hardly get near the tree. We use these seedlings to graft the thornless cultivars on.

CULTIVARS: 'Rubylace'. This tree is the same in all respects except that the leaves are a ruby colour. An unusual colour to get in a leaf but very beautiful. No thorns on this one, either.

'Shademaster'. What an accurate name for a tree. This one again is very similar to the other two but has green leaves that produce good shade.

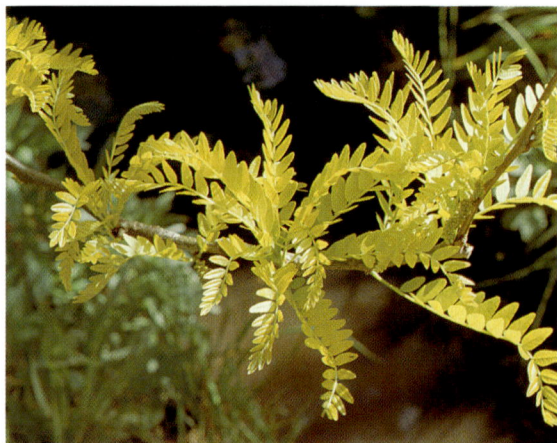

GOLDEN CHAIN

(Laburnum x watereri 'Vossii')

FEATURES: Deciduous; height to 6–7 m; spread to 2–3 m. Non-invasive.

FLOWERS: Pea-shaped, hanging in chains up to 40 cm long. A faint perfume to these golden chains. Spring-flowering.

LEAVES: Trifoliate, looking for all the world like clover leaves. The leaves are also furry with little silver hairs all over them, giving a glistening silvery hue when the dew is on them.

WHERE: Prefers a cooler part of the garden, sheltered from strong winds. A cool root system in rich soil will make the laburnum perform well. Deep mulch over summer and water well.

USES: This is a tall, narrow tree that maintains a green trunk and branches for many years. It is used primarily as an ornamental. In Bodnant in Wales you will find one of the world's most famous botanical sights — a laburnum walk. A tunnel of poles has had a laburnum planted on each pole and pulled over the rounded top so that they meet — the tunnel is now made up of only laburnums. When they flower all the chains of golden flowers hang through the inside of the tunnel, an unbelievable sight!

COMMENTS: Originally from Scotland, the laburnum has featured in cottage gardens for years. The seeds are poisonous but, like so many poisonous things, are extremely bitter (I found out from experience, and am still here to tell the story). When buying one of these fellas, make sure it is grafted. That is why its botanical name is so long, it is a cross between two types of laburnum.

GOLDEN ELM

(*Ulmus procera* 'Louis van Houtte')

FEATURES: Deciduous; height 8–10 m; spread 5–6 m; needs hand watering over the first two summers to encourage deep roots.

FLOWERS: Insignificant.

LEAVES: Large, furrowed and golden. They cover the tree in an upper canopy. Starting a rich gold in spring, changing to a soft, golden green in summer and then to an old gold in autumn. One of the best foliage trees.

WHERE: Elms can take a year to settle in and then make their move with fast growth over the next few years, before slowing right down, at which point they are a respectable beautiful tree. Not happy by the coast but the middle coastal plain is fine, and up in the hills is best. Elms require a good watering over summer until established.

USES: Elms provide marvellous shade and can be pruned to 2 m to allow the canopy to grow over. If you have room for one spreading tree, this could be it. May sucker if roots are damaged.

COMMENTS: There is nothing quite as awesome as a Golden Elm in full spring or autumnal colour. Their habit of sulking for a year after planting is more than made up for by their subsequent growth.

CULTIVARS: 'Silver' Elm has a variegation in the leaves that makes them appear silver all over. A more upright-growing form, and fast when it takes off.

The 'Fastigiate' or 'Upright' Elm has remarkable leaves that wrap around the stems. A very narrow and smaller elm that can be grown in a metre-wide bed. They look wonderful in a row instead of bloody pencil pines.

The 'Weeping' Elm is grafted on a 2 m trunk and then grows outwards and downwards. An ideal specimen tree that is rare and fascinating. They look marvellous over a pond or running water.

VARIETIES: The Chinese Elm (*Ulmus parvifolia*) has quite small leaves, grows well from the coast all through to the hills and produces an upper canopy offering very good shade over summer. Dark green leaves. This tree can be shaped to a form of topiary on a grander scale.

HANDY HINT

WARMTH-LOVING TREES. The subtropical trees that flower so gloriously over summer love a warm spot in your garden, both summer and winter. Among these are the Royal Poinciana, Leopard Tree, cassias and the lovely Golden Poinciana. High humidity is one of their requirements and there is a very simple way to help boost this in our hot, dry summers. Spread a bale of hay around the base of the tree over your deep mulch. Make sure this is kept very moist and as the sun hits it in the morning it will start to steam. The resultant moisture will rise to the underside of the leaves, giving high humidity in a close confine. This will open the stomata and help stimulate growth.

KEEP FREE
100 mm

MOISTURE RISING
IN HEAT

DEEP MULCH

STRAW ON TOP
OF MULCH

GOLDEN POINCIANA

(*Schizolobium parahybum*)

FEATURES: Deciduous to semi-deciduous; height to 10 m; spread to 8 m. Non-invasive.

FLOWERS: Very similar to the Royal Poinciana, only golden. It belongs to the same family as the poinciana and flowers at the same time. The flowers are carried across the upper canopy.

LEAVES: Remarkable. They are very long (up to 2 m) with many paired leaflets. They grow out and downwards like a frond, producing loose shade when young and more dense as the tree matures.

WHERE: Faster growing than the Royal Poinciana, they will grow anywhere in Perth, even over limestone. They love a rich soil and good watering while they are growing, over the hot months. They are more tolerant of cold than the poinciana and look good in the hills.

USES: If allowed to reach their full height and you have heavy winds, they can be prone to branch damage. Therefore, best to cut them off at about 5 m and let them branch. Ideally planted as a very showy specimen tree.

COMMENTS: The up side of this tree is that it will flower in 4 or 5 years, or possibly sooner. Rarely seen in Perth and I don't know why, because they thrive here. If you are after a tropical look, seriously consider this tree.

GOLDEN RAIN

(Koelreuteria paniculata)

FEATURES:
Deciduous; height to 5-6 m; spread to 2-4 m. Non-invasive.

FLOWERS: Small, golden, carried on the outer and upper edge of the tree. Late spring to early summer flowering. A pretty tree with the mix of gold flowers against a dark green backdrop.

LEAVES: Indented on the margins, oval to elongated in shape, mostly carried on the perimeter of the tree. They turn a buttery yellow in autumn.

WHERE: Anywhere except right on the coast. High winds can damage the leaves. Adaptable to all soils, it likes a good deep mulch and regular fertilising.

USES: Can provide excellent shade if trained up to allow a canopy to develop. Good for shading azaleas and the like as an acidic soil doesn't bother it.

COMMENTS: This is one of the trees recommended by the American Tree Society as a beautiful garden tree. Good under power lines. The seed pods are a special feature and can be used in dried flower arrangements.

GOLDEN SHOWER

(Cassia fistula)

FEATURES: Evergreen or semi-deciduous; height to 5 m; spread to 4 m. Non-invasive.

FLOWERS: A pure yellow pea flower, hanging in clusters or bunches like grapes. Flowering can commence after Christmas and continue for many months. The tree can be smothered in blossom and it really does look like a golden shower.

LEAVES: Sea-green, large and similar to eucalypt leaves, hanging downwards. New growth is a coppery-bronze colour. If the winter is mild or warm the tree will retain all its leaves as it does in the tropics; but if the winter is cold it will shed completely.

WHERE: Fine on the coast and the coastal plain. I advise planting an advanced tree in the hills. The larger the tree, the less problem with frost. Highly adaptable in its soil requirements, it does best in a fertile soil with good hand-watering over the hot months.

USES: Mostly ornamental. Unpruned, this tree is almost as broad as it is high and will serve as a screen. Good on a west- or north-facing fence.

COMMENTS: Most good nurseries will have plenty of stock on hand when this tree flowers. The phones run hot with inquiries from people who have spotted one in flower. Many Golden Showers are grown in Victoria Park — worth a drive around there after Christmas to see this glorious tree flowering.

VARIETIES: *Cassia javanica*, also known as the Apple-blossom Cassia because of its vibrant pink and white flowers. The leaves are paired and it, too, has a spreading habit.

Cassia nodosa has pink flowers. Again a spreading tree with a most gorgeous show.

HAWTHORN

(Crataegus monogyna)

FEATURES: Deciduous; height to 5 m; spread to 2 m. Non-invasive.

FLOWERS: In a broad panicle of many hundreds of pure white, perfumed flowers. These give way to pretty red berries that are said to be a cure for arthritis. The hawthorn belongs to the rose family.

LEAVES: I love the leaves of the hawthorn. They are indented and a little like an hourglass in shape (this is not a good description; you really need to see them). The autumn colour is glorious, deep scarlets and reds with the odd burnt orange leaf.

WHERE: Good on the coastal plain and great in the hills. A rich, well-drained, moist soil will show them off.

USES: The hawthorn has been used in England for centuries in hedgerows, and for good reason. They branch from the ground and make a hedge no-one can get through. The thorns are wicked, up to 40 mm long and very sharp. They can be cut off, particularly from the lower branches. Ideal if you are worried about burglars. They also keep animals out, or in.

COMMENTS: Hawthorns are most attractive in their own right; whether you use them for a hedge or not, one on its own is fabulous.

CULTIVAR: *Crataegus laevigata* 'Paul's Scarlet' is a thornless form with rosy red flowers, otherwise much the same habit.

ILLAWARRA FLAME TREE

(Brachychiton acerifolia)

FEATURES: Evergreen to semi-deciduous; height to 8 m; spread to 5 m. Non-invasive.

FLOWERS: Scarlet bells that hang in profusion from this unusual tree. Flame Tree indeed! It certainly looks as if the whole tree is on fire when in flower. There is a glorious specimen in St George's College in Crawley that has to be protected from souvenir hunters.

LEAVES: 'Acerifolium' means a leaf like a maple (*Acer*) because it is lobed or hand-shaped. Very large and soft green, the leaves make the tree very handsome when out of flower. When they drop it seems to be at any time other than autumn and is prior to flowering.

WHERE: Hailing from Queensland and New South Wales, the Flame Tree fits into Perth very well. It seems to suffer from very little and tolerates harsh conditions. A small amount of fertiliser is sufficient for a long time. Coast to hills.

USES: As a specimen tree for foliage and flowers, although the boat-shaped seed pods are quite unusual, too!

COMMENTS: There is a line of Flame Trees in Midland between the highway and the crescent backing onto a park. Tourist buses detour that way in summer when the whole row is in flower (except one that flowers a month later!). The road is also covered in the vivid blossom. Some trees develop a lop-sided form.

VARIETIES: The Kurrajong (*Brachychiton populneus*) is a more upright form with soft pink and green bells. The leaves are much smaller and a darker green.

JACARANDA

(Jacaranda mimosaefolia)

FEATURES: Semi-deciduous or deciduous; height to 15 m; spread to 12 m. Needs watering by hand for at least two summers to establish roots.

FLOWERS: Purple/mauve, trumpet-shaped, about 40 mm long. The spectacle of a jacaranda in full bloom is worthy of poetry.

LEAVES: Very fine, with paired leaflets and appear to be quite delicate. The tree gives good dappled shade. Leaves drop prior to flowering, briefly.

WHERE: The coastline itself is not good for these trees, but from a couple of kilometres back through to the hills is fine. In the hills, care must be taken to establish the tree before the cold hits it and burns the foliage back.

USES: As a parkland tree, street tree or on a large property. A stroll through Hyde Park in North Perth when the jacarandas are blooming is unforgettable. South Perth is bestrewn with them.

Well-established trees usually have their roots everywhere but where you want them, and although the dappled shade is great for growing plants under them, the roots often are too prolific. Hence the need for deep-watering young trees!

COMMENTS: When used in an avenue, as in some streets in Mount Lawley, it creates a leafy glade under which we would all love to live. When it falls, the blossom literally carpets the ground where it is just as spectacular.

CULTIVARS: The 'White' Jacaranda is a grafted tree growing to around 5 m. The flowers are pure white.

The 'Variegated' Jacaranda has an unusual variegation in the leaf. It looks as though a broad brush has been wiped across the leaves with a yellowy/white colour. Grows to around 5 m.

JAPANESE MAPLE

(Acer palmatum)

FEATURES: Deciduous; height to 5 m; spread to 3 m. Non-invasive.

FLOWERS: Very small. You have to hunt for them, usually red, quickly followed by the winged seeds or samaras.

LEAVES: Now this is where I could go feral (as the kids say!). The second part of the botanical name means palm (as in hand) -shaped. Mostly the Japanese Maples have five or seven 'fingers' to the leaf. The tree typically known as Japanese Maple is a very lovely small tree with the leaves as described, but there are over 450 cultivars of this plant and each has a variation in the leaf. It may be very finely dissected, bubbled like silver beet, coloured purple, variegated with a lot of white and green, variegated with one stripe of white on a green leaf, variegated with white, pink and green, and on and on. Those who collect them get very excited about a new discovery, but my experience has been that only 40 or so cultivars will look good in our climate.

WHERE: All the cultivars can be grown in containers and kept in a sunny but wind-free area. Once the wind gets to them it desiccates the leaves. The tree Japanese Maple can be grown in a sheltered part of the garden in full sun. The middle coastal plain and hills would be best.

USES: Just to look at and marvel at the beauty.

COMMENTS: The whole range of Japanese Maples will colour brilliantly in autumn. Even the bonsai trees colour well. The secret is to slow down the watering a little in autumn and stress them slightly. There is a bit of an art to this, but it is worth the patience.

JUDAS TREE

(*Cercis siliquastrum*) also known as Redbud

FEATURES: Deciduous; height to 4–5 m; spread to 2–2.5 m. Non-invasive.

FLOWERS: Hot-pink pea flowers right on the branch or trunk. They appear prior to leaf set in late winter and fairly glow in the overcast light. One of the best trees to herald the end of winter.

LEAVES: An oval heart shape, most attractive and quite sturdy. Autumn colour is nothing special.

WHERE: The Judas Tree is a lime-tolerant species and does well by the coast. As it is a small tree the wind doesn't do much damage. Equally at home in the hills and all points between.

USES: Can be grown as a hedge by keeping it clipped to shape, otherwise as an ornamental. A rather nice way of growing the Judas Tree is to plant a white- and a pink-flowering form in the same hole. They will grow up together and flower at the same time creating interesting comment!

COMMENTS: The Judas Tree has been thought of as the tree upon which Judas hanged himself. Not so! The name comes from the French 'l'arbre de judah' meaning 'tree of judah'.

VARIETIES: *Cercis canadensis* has white-flowers but is the same in all other respects.

KAFFIR PLUM

(Harpephyllum caffrum)

FEATURES: Evergreen; height to 7–8 m; spread to 4–5 m. Non-invasive.

FLOWERS: Greenish-white, not spectacular, in early summer.

LEAVES: Tough, leathery, dark green, nicely carried on a spreading tree.

WHERE: This tree has proven itself able to grow anywhere. A native of South Africa, it enjoys our climate and conditions as if it were home. If grown by the coast it's worth hosing it down every now and then to remove salt-spray.

USES: Great for hiding two-storey extensions or nosy neighbours. A good wind-break. The plums that are produced make a good jelly.

COMMENTS: A good tree to grow for the black-thumbed gardener. It is hard to go wrong with this one.

HANDY HINT

FLIES. A simple way to help lower the fly and blowfly population is to take a clear-plastic, two litre soft drink bottle and, about two-thirds the way up, make a hole big enough for flies to enter. Unscrew the lid and drop some mince, a small handful, and a cup of water in the bottom. Screw the lid back on and hang the bottle up away from the house. The mince will rot, and whilst the flies can smell it, you won't. They will be attracted to the bottle, climb through the hole and bingo! They will try and escape via the lid and keep falling to the bottom until they drown. At the end of summer throw the whole bottle away.

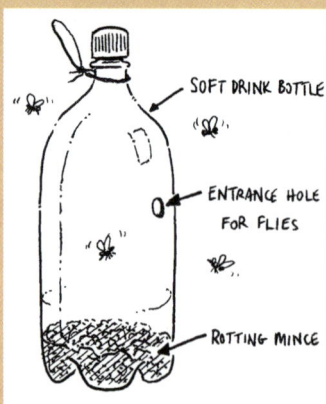

SOFT DRINK BOTTLE

ENTRANCE HOLE FOR FLIES

ROTTING MINCE

KILMARNOCK WILLOW

(*Salix caprea* 'Kilmarnock')

FEATURES: Deciduous; height to 3 m; spread to 1.5 m. Non-invasive.

FLOWERS: Small furry 'pussies' as this is a form of the pussy willow. Silvery-white, they are very effective, if unusual.

LEAVES: Small, oval and leathery, with a textured surface. They are a dark green, turning to a real buttery yellow in spring. As this is a weeping tree, the leaves cascade down the branches.

WHERE: Fine on the coastal plain and in the hills. Typically the willow likes a good watering and will benefit from an enriched soil. Once established it will withstand most of the elements, though it would be wise to avoid fierce winds.

USES: This is a very small willow, graceful and a splendid specimen tree. Ideal for courtyards or confined areas. Very architectural in its appearance.

COMMENTS: This is the willow to grow if you want one that is petite. It is very well behaved in any garden, rarely, if ever, requiring pruning. When it is mature the weeping branches will reach the ground, looking like a waterfall of leaves.

VARIETIES: The Contorted or Tortured Willow (*Salix matsudana* 'Tortuosa') is a small tree with all its branches, even the trunk, twisted and contorted. This is the one you see in dried flower arrangements.

New Zealand Hybrid Willow is the only drought-tolerant willow. It grows in an upright, spreading way and is a very attractive small willow.

The Cricket Bat Willow (*Salix alba* var. *caerulea*) is the one all small boys would like to have. You can give it to them, it won't grow too large or unwieldy.

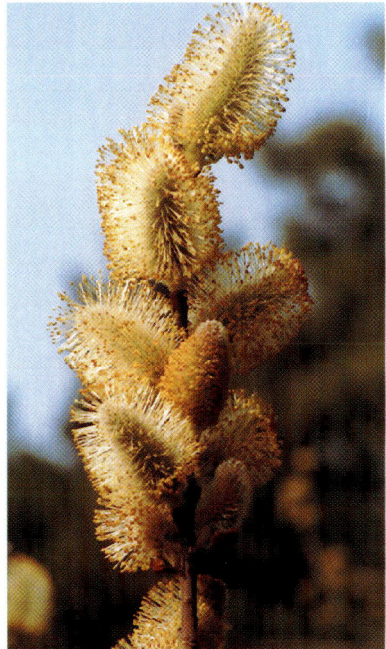

LAGERSTROEMIA

(*Lagerstroemia indica*) also known as Crepe Myrtle

FEATURES: Deciduous; height to 5–6 m; spread to 4 m. Non-invasive.

FLOWERS: In clusters on the end of the branches. Close inspection reveals a creped-ruffled edge to the petals. The colours come in pure white through to deepest scarlet and all shades between.

LEAVES: Generally oval, about the size of a ten cent coin. Autumn colour can be quite magic with deep scarlet leaves.

WHERE: Lagerstroemias like a warm to hot spot in the garden. They can be grown everywhere except the actual coast. They love the hills and are just as happy in Perth city, where you will find some gorgeous trees in the Old Mint front garden. Once established, they can be drought-resistant.

USES: Primarily as a specimen tree. They do make a good screen, and the dwarf varieties make an excellent hedge.

COMMENTS: Long flowering; you will see them from late summer all the way through to the end of autumn. Something else rather special about these marvellous trees is the shiny trunk they develop with age. They can suffer from mildew, although I have never experienced that problem.

LIQUIDAMBAR

(Liquidambar styraciflua)

FEATURES: Deciduous; height to 15 m; spread to 5 m. Non-invasive.

FLOWERS: Insignificant.

LEAVES: Very similar to a maple with dissected leaves, dark green from spring to summer. Autumn colour is spectacular — the whole spectrum from yellow to blood red. Very attractive tree because of its conical shape (typically, it grows to a point) and its coverage of leaves.

WHERE: You will find liquidambars everywhere in Perth. They seem to have very little in the way of special soil requirements, but if you want a really beautiful tree give it rich soil, regular fertiliser and good watering.

USES: Good for hiding things, but really a specimen tree. They will retain branches from the ground up if you don't prune them off. I like them that way.

COMMENTS: Actually a member of the witch-hazel family, they can be slow-growing in ordinary soil. They look awful if the top is cut out.

CULTIVARS: 'Jennifer Carroll' is a miniature form growing to about 2 m. Identical to the normal tree but tiny.

'Palo Alto' has blood-red leaves for about six weeks and grows to only around 7 m.

'Canberra Gem' is a cultivar chosen for its excellent colours; grows to 8 m.

VARIETIES: The Variegated Liquidambar has green and white leaves, and in autumn pink is added to make three colours.

LONDON PLANE

(Platanus x acerifolia)

FEATURES: Deciduous; height to 10–15 m; spread to 10–15 m. Has invasive roots.

FLOWERS: Insignificant.

LEAVES: Large and broad, slightly dissected, coming to a point. Hairy and a wonderful soft green. The shade cast by a London Plane is almost translucent. When the leaves fall they can create large drifts and it is always tempting to roll in them.

WHERE: Easy to grow, but hard to keep the roots under control. When they are watered well from first planting, the roots will grow deeper, but it is against the nature of the tree to do so. They thrive on the coastal plain (there are many planted along The Esplanade in the city) right through to the hills. Guildford Grammar School has an avenue of mature specimens with some younger trees to complete the effect.

USES: Planted in the right spot, a plane will produce a canopy and shade for a huge picnic for all the family! Lining either side of a long driveway, they will create a tunnel under which to drive or walk — but you need acreage.

COMMENTS: Although the autumn colour is ordinary, I love driving past Guildford Grammar when the road is covered by plane leaves and watching them fly up behind the car. One way to keep the tree reduced in size is to pollard it. This means removing all the branches at about 2 m and letting the tree shoot again, then repeat the process next year. Best done in winter.

VARIETIES: The Oriental Plane (*P. orientalis*) is a slightly smaller plane with more dissected leaves. It doesn't produce the mottled colour on the bark as does the London Plane but is still a lovely tree.

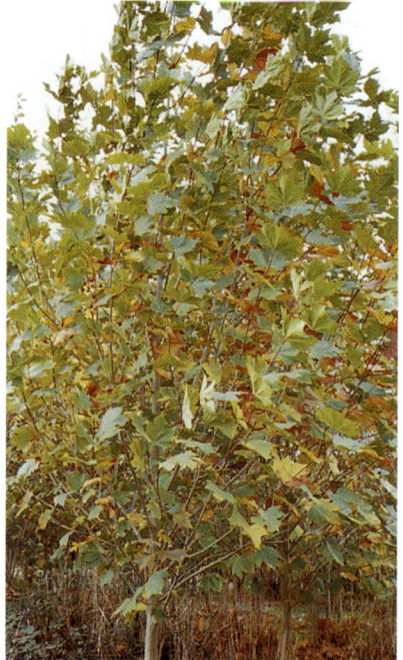

MAGNOLIA

(Magnolia x soulangiana)

FEATURES: Deciduous; height 3–4 m; spread 2 m. Non-invasive.

FLOWERS: Commonly up to 25 cm across and shaped like a large goblet. A large flower by any standards, they are mostly white with a flush of red at the base of the large petals. Spring- to early summer-flowering, and sometimes a second flush (though not as showy) in early autumn.

LEAVES: Long, broadening towards the end, then coming to a narrow point at the tip. The leaves often turn a coppery/magenta colour if autumn is cold. Will offer some shade.

WHERE: If you can avoid strong, hot winds, the magnolia will do well on the coastal plain and in the hills. A deep mulch and rich soil is a basic requirement, but the most important ingredient is water; don't let them dry out when in leaf.

COMMENTS: Just after Hyde Park and before Fitzgerald Street on Vincent Street are two deciduous magnolias about 3 m tall. Each is a mass of blossom in spring. There are many cultivars of the deciduous magnolia with some very distinctive flower shapes.

CULTIVARS: 'Heaven Scent' has dark pink, elliptic petals, quite a large flower with a strong fragrance.

VARIETIES: *Magnolia stellata* (star-shaped) has pure white petals that fall outward as the flower opens, a very showy small tree.

MAIDENHAIR TREE

(Gingko biloba)

FEATURES: Deciduous; height to 10 m; spread variable, from upright to 10 m. Non-invasive.

FLOWERS: This is an amazing tree which can be either male or female and can take up to 15–20 years to become sexually mature. The 'flower' itself is not your normal run of the mill either. The tree is classified as a gymnosperm — which nearly puts it into the conifer family, but not quite. The male produces a motile sperm that has to move from the male tree to the female in order to complete the union. But wait, there's more! Once fertilised, the female then produces a plum-like fruit with a kernel inside. The fruit has a vile smell and has caused some concern among potential purchasers of the tree. However, the chances of you having a 20-year-old female with a male within cooee are so remote that there is a better chance of politicians telling the truth!

LEAVES: Well, the amazing bit continues. The leaf is shaped just like a maidenhair fern leaf, only about the size of a fifty cent piece. It is quite leathery and, as autumn commences, starts to go a soft yellow around the outer edge. Slowly the whole leaf turns gold as the tree removes the chlorophyll. One of the most vibrant of autumn colours is a gingko in full flight.

WHERE: The story continues. As it is not overly concerned with soil type, you can grow the gingko almost anywhere in Perth. Strong wind can burn the leaves, so shelter from violent wind is good. A little fertiliser is appreciated, but don't go overboard. I have seen a large, good-looking tree grown in the heart of New York City where one paving slab was lifted to plant the tree. Heaven only knows how it survived the fierce cold, the intense heat and the smog — but survive it did.

USES: Just grow it for its beauty and uniqueness.

COMMENTS: The gingko hasn't changed since the days of the dinosaurs. Fossils of the gingko 200-million-years-old are the same as the tree of today. Maybe that's why nothing bothers it. It was thought to be well and truly extinct until some botanical explorers found a group growing in China in the 1920s.

I kind of like the idea of growing a living fossil in my garden.

HANDY HINT

ON TREE PLANTING HOLES. When planting trees in heavy soil, like the Darling Range soils, you can use an auger or post-hole digger, but make sure you make the hole square with a crow-bar, and loosen up the soil at the bottom of the hole. This way the trees roots can make a grip when growing rather than spiralling. If you don't, the trees may make good initial growth but after a few years just blow over.

TREE HOLE

POST-HOLE DIGGER

CROW-BAR

SQUARE HOLE FIRST

MANCHURIAN PEAR

(Pyrus ussuriensis)

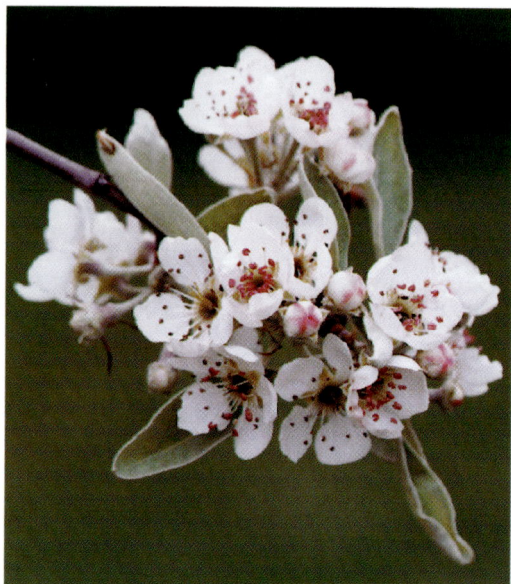

FEATURES: Deciduous; height to 6 m; spread to 3 m. Non-invasive.

FLOWERS: Small, single, pure white, in spring. The tree is covered with them and they look quite dainty.

LEAVES: Typical pear, glossy and dark green. In autumn, changing first to deep oranges and then vibrant reds. Good shade and lush over summer for a cooling effect.

WHERE: Pears are very tough trees, fast growing and tolerant of most soil types. If you can give it rich soil, a good mulch and water it will perform fabulously. Coastal plain through to the hills.

USES: Ornamental as it doesn't produce fruit. Can be interwoven very effectively and espaliered.

COMMENTS: We have one by our front door. It provides shelter, shade, beauty and a wealth of autumnal colour. It is growing in a tough, rocky, dry part of our garden and thrives on neglect.

CULTIVARS: 'Winterglow' does all of the above, but is out of leaf for only about six weeks instead of three months.

'Totem Pole' pear grows as a column to 5 or 6 m and only 1 m wide. In spring this column is a mass of pure white blossom, and afterwards clothed in leaves. Will grow in a bed 1 m wide. A row of them looks superb, much better than bloody pencil pines.

VARIETIES: The Willow-leaf or Silver Pear (*P. salicifolia* 'Pendula') is a weeping form with narrow, silvery leaves. It has white flowers and is at home in a cottage garden. Absolutely stunning.

NATIVE FRANGIPANI

(Hymenosporum flavum)

FEATURES: Evergreen; height to 10 m; spread to 4 m. Non-invasive.

FLOWERS: Small, yellowish, in profusion over a very long period from early summer to late summer. Very fragrant, and very pretty to see the tree studded in these flowers.

LEAVES: Small, dark green, oval to elliptic in shape. Quite leafy although the Native Frangipani offers only moderate shade.

WHERE: The tree grows quite happily anywhere in Perth from coast to hills. It is not fussy about soil type and is very easy to grow. It takes strong winds with one exception: it can be brittle and lose a branch in a howling gale.

USES: If kept well trimmed it can make a good narrow screen to fit into a tight area. Otherwise just as a shade and specimen tree.

COMMENTS: We have a large tree in our garden and it always catches me out. I keep looking for the flowers, and the day I don't — there they are! This tree is an absolute picture for many months, and so undemanding. Great for a beginner gardener.

HANDY HINT

BULBS. Prior to planting out bulbs for spring flowering, in particular tulips and daffodils, place them in a paper bag and leave them in the crisper of your fridge for a month or so. The best time to plant is toward the end of May, so just pop them in at the end of April and the show will be much better, as they enjoy being chilled. To choose from the many, many varieties of tulips and daffs I recommend a visit to Araluen in spring to see the magnificent bulb display they have there every year. As they flower over many weeks you won't miss out and will be able to see them in the flesh which is far better than any photo.

NEEM TREE

(Azadirachta indica)

FEATURES: Evergreen; semi-deciduous; height to 5 m; spread to 3 m. Non-invasive.

FLOWERS: Small, pretty, white and mauve, the size of a five cent coin, in late spring and early summer.

LEAVES: Small, narrow, indented or serrated, dark green. Makes a good dappled shade tree.

WHERE: A warm spot in your garden both summer and winter. It will grow happily on the coastal plain. Requires good watering when young. Protect from frost in the hills.

USES: This is the tree that the United States is trying to corner the market on. Why? Because it produces the strongest, naturally occurring pesticide in the world. Needless to say the tree suffers no damage from insects! Oil is extracted from the berries that are produced after flowering. To make your own organic pesticide, collect the berries and crush them. Place in a blender with water and blend. Steep the solution in a bucket of water for a week or so, draw off that and dilute 100 mls to a litre.

COMMENTS: Not only an attractive tree, but how useful! A good tree for schools to grow and experiment with the berries.

NETTLE TREE

(Celtis australis)

FEATURES: Deciduous; height to 5 m; spread to 3 m. Non-invasive.

FLOWERS: Insignificant.

LEAVES: Long and oval, serrated. The autumn colour is quite unusual — a soft pinkish grey turning yellow.

WHERE: Best in the hills where it is cooler in autumn. Likes the gravelly soil of the hills. Easy to grow.

USES: Being a member of the elm family, it will cast good shade, although it is quite small. Best grown for its autumn colour.

COMMENTS: Late autumn, when the evenings come in sooner and dusk settles, is when this tree glows. The colour is magical and very beautiful.

HANDY HINT

POTTING. These days terracotta look-alikes are so similar to the real thing that you have to get pretty close to pick the difference. There is, however, one sure distinguishing point. They can't put the drainage holes in the bottom — you have to do that. If you don't, you can kiss your plants bye-bye because they will drown. So arm yourself with an electric drill or a brace and bit (you'll know what that is if you're my age) and bore about five or six holes through the bottom. I recommend at least a half inch (1 cm) bit to make sure the holes are big enough.

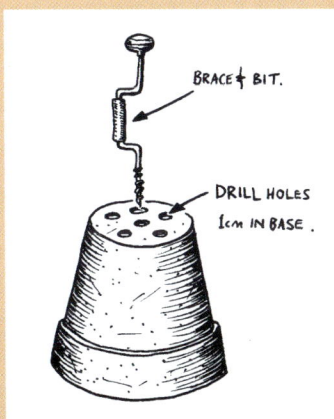

BRACE & BIT.

DRILL HOLES 1cm IN BASE.

PERSIAN WITCH-HAZEL

(Parrotia persica)

FEATURES: Deciduous; height to 6 m; spread to 5 m. Non-invasive.

FLOWERS: Insignificant.

LEAVES: Elliptic, about 60 mm long. They hang overlapping on this small tree. The effect is an outer covering of leaves. The autumn colour is very rich in the red hues, soft green beforehand.

WHERE: A protected area away from strong winds, in a quiet spot in the garden. They will take dappled sunlight or part shade and love a rich soil, deep mulch and good watering when young. They are quite drought-tolerant when older, but will look a little sparser. Okay on the coastal plain, better in the hills.

USES: Mainly as an unusual small tree that will intrigue your visitors. When it colours in autumn it will stop them in their tracks.

COMMENTS: From the same family as the Liquidambar which is also very special in autumn. The beauty of the Parrotia is its ultimate weeping habit, with the leaves looking like water cascading down the outside.

HANDY HINT

LIVER FOR PASSIONFRUIT? I was once advised by an Italian lady to place a pig's liver under a passionfruit. Being somewhat impatient in those days, I put four underneath an ailing passionfruit vine. The vine went berserk and eventually brought down the pergola it was growing on. This pergola was built by me, which says a lot for my carpentry skills! The fruit we obtained from this vine would have been enough to set up a roadside stall. I think any liver would do and my reasoning is that liver contains large amounts of iron — one of the more important elements required by plants. Try it — but check your pergola first!

PIN OAK

(Quercus palustris)

FEATURES: Deciduous; height to 10 m; spread to 4 m. Non-invasive.

FLOWERS: Inconspicuous.

LEAVES: Large, deeply cut or lobed, a rich, shiny green from spring onwards and turning crimson and scarlet in autumn. They then dry and remain on the tree through winter. The leaves are quite unusual and give the tree an intriguing look.

WHERE: Pin Oaks are grown extensively in Canberra, Melbourne and Sydney, in three differing climates, and they thrive in each. Preferring a rich soil and good water, they will grow well in the middle suburbs and up into the hills. Near the new public library in Midland is a street planting of Pin Oaks and they are doing well in the heavy clay of that area. They don't appear to need any special treatment.

COMMENTS: One of the more spectacular of 350 or so oaks that you could plant. By far a smaller version of the English Oak (*Quercus robur*). The autumn colour on such unusual leaves just adds to their beauty.

VARITIES: Cork Oak (*Quercus suber*) is an evergreen that is used commercially to produce cork. Small holly-like leaves. Very tough. There are good examples at The University of Western Australia.

Cypress Oak (*Quercus robur* 'Fastigiata') has a narrow form that is a nice substitute for bloody pencil pines.

The Golden Oak (*Quercus robur* 'Concordia') — if you can find one — is a golden version of the English Oak, only much smaller. Its spring and autumn colours are golden and vibrant. Worth the hunt!

Willow Oak (*Quercus phellos*) is a small, rounded tree with willow-like leaves. Just gorgeous.

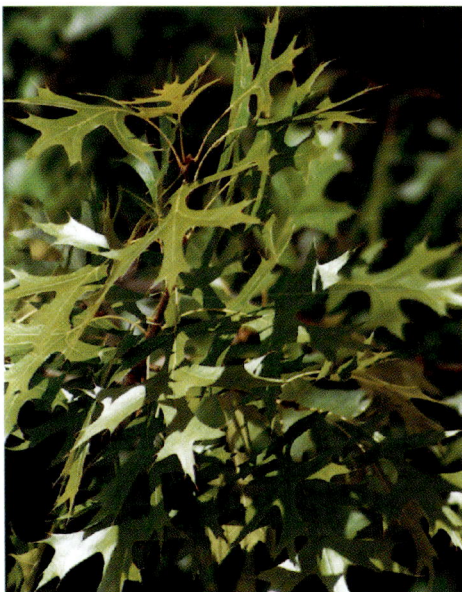

PINK WISTERIA TREE

(*Robinia pseudoacacia* 'Decaisneana')

FEATURES: Deciduous; height to 8 m; spread to 4 m. Requires deep watering to encourage a deep root system.

FLOWERS: Highly perfumed bunches of soft, rich pink pea flowers. They hang on the tree like wisteria, hence the name.

LEAVES: Slightly elongated, oval, green from spring onwards, turning a buttery yellow in autumn. Not dense but gives reasonable shade.

WHERE: Will grow anywhere in Perth. It is damaged by fierce winds but will tolerate average summer winds. It can have brittle branches. Very fast growing, it likes a manure-rich soil and good watering.

USES: Not a particularly large tree, ideal as a specimen for the showy flowers. Keep it near the house to enjoy the heady fragrance.

COMMENTS: When I close my eyes and inhale the perfume of this tree, I imagine I'm in the West Indies or the Caribbean — a very cheap way to travel! The bunches of flowers are a sight to behold.

CULTIVARS: 'Frisia', first discovered in a French nursery last century. The tree is the same as above, except that the leaves are golden. The flowers are a golden-white and perfumed.

'Mop-top Robinia'. This is one of the most popular trees of all. Grafted usually at 2 m, it develops a round head 2 m across which is a mass of deep green oval leaves. It forms this head all by itself — no cutting or pruning! Rarely if ever flowers.

VARIETIES: Weeping white wisteria tree (*R. uniflora*). This tree has an elongated leaf and a cascading habit with its branches. The flowers are like a large prominent bunch of white wisteria flowers.

POWTON

(*Paulownia fortuneii*); Powton Sapphire Dragon (*Paulownia kawakami*)

FEATURES: Deciduous; height 8–10 m; spread to 4–6 m. Non-invasive.

FLOWERS: Soft mauve fading to pinkish white with a speckled yellow and brown throat. The Powton Sapphire Dragon is a blue flower, again with a speckled throat. In both the flowers are trumpet-shaped, fragrant, and last for six weeks; the buds are formed in autumn but don't open until spring. Flowering is usually in the second year.

LEAVES: The leaves of both trees are huge at the juvenile stage – you can't get your arms around them. As the trees mature, they reduce to about one quarter of that size. The shade given by both trees is excellent, without preventing other plants from growing underneath. It's worth collecting the leaves in autumn as they make the best compost.

WHERE: Powtons are now growing all over Perth, from the coast through to the hills, and thriving. They have a deep tap root that taps into available minerals and puts them back into the garden at leaf drop. For the first three years they grow extremely quickly, up to four metres in the first year, and then slow right down. I often call them the '$20 pergola'. Plant one in the backyard and put a sandpit under it for the kids; they will have all the sun protection they need.

USES: The broad canopy will reduce the summer temperature by five degrees. This is one of the new generation timber trees to replace the hardwoods currently being imported from rainforest areas. The powton offers a good screen and windbreak.

COMMENTS: The powton is native to China, where it has been cultivated for thousands of years. There are many varieties, but these two have proven themselves to be the best for WA conditions. If you buy a powton, make sure it has the red or blue powton label; there are a few look-alikes!

PRAIRIE CRAB-APPLE

(Malus ioensis 'Plena')

FEATURES: Deciduous; height to 3 m; spread to 1.5m. Non-invasive.

FLOWERS: Cup-shaped, soft pink with an almost apricot blush. Flowers appear in spring in clusters and are double. The texture of these flowers makes you think they are made of porcelain.

LEAVES: Large and oval-shaped with a dissected edge. Surprisingly large, when you consider the delicate flowers. Quite good autumn colour in burnt oranges and reds but not prolonged.

WHERE: In a protected area away from violent winds. Not as tough as smaller-leaved crab-apples, but stunning if you have the right spot. Fine from the coast through to the hills in a rich soil with good watering.

USES: The vase shape of the tree makes it quite compact and very neat in appearance. Best used as a feature ornamental. Will cast some shade.

COMMENTS: Although it may be difficult to find, it is finally available in Perth. Anyone who has seen this little tree in flower wants one. It is my favourite crab-apple, and well worth finding, firstly the right spot, and secondly the tree. There is a specimen among the crab-apple collection at Araluen.

Typically, as crab-apples were an essential part of cottage gardens, they will all fit in and are commonly grown as a group.

CULTIVARS: 'Purpurea' has pink to purple flowers and small red fruit. A very upright form with dark leaves.

'Echtermeyer' is a very special weeping form. The flowers look like hand-painted porcelain with a white background and then a brush-stroke of mauve on each petal.

'Golden Hornet' produces gold crab-apples that linger on the tree.

ROYAL POINCIANA

(Delonix regia)

FEATURES: Deciduous; height to 6–7 m; spread to 8–10 m. Non-invasive.

FLOWERS: Each flower is as intricate as an orchid and just as beautiful. The colour is deep orange/red to scarlet. The flowers are tubular opening outwards to reveal the stamens. When in flower, generally mid- to late-summer, the tree can be covered over the entire canopy with these flowers.

LEAVES: Bipinnate, with small, soft paired leaflets about the size of a little fingernail. The colour is a light green and the appearance lush. They cover the canopy of the tree and provide dappled shade.

WHERE: Fine on the coastal plain and up to 1–2 kilometres from the beach. They are tolerant of all soil types but thrive in a very rich, moist soil. Poincianas love heat, both summer and winter, so plant in the hottest part of your garden. In the hills your only hope is with an advanced specimen planted in spring and protected in winter with a cover. The same advice for spots subject to frost on the coastal plain.

USES: The shade is fabulous and smaller sub-tropical shrubs look at home growing under them. If necessary, the tree can be pruned heavily to contain the spread. A glorious tree for its foliage, and as for the flowers — just nothing compares.

COMMENTS: Justifiably argued as the most beautiful flowering tree in the world, the poinciana has one down side. It can take up to 15 or more years to flower here. Fortunately the foliage is so appealing! I would still plant one.

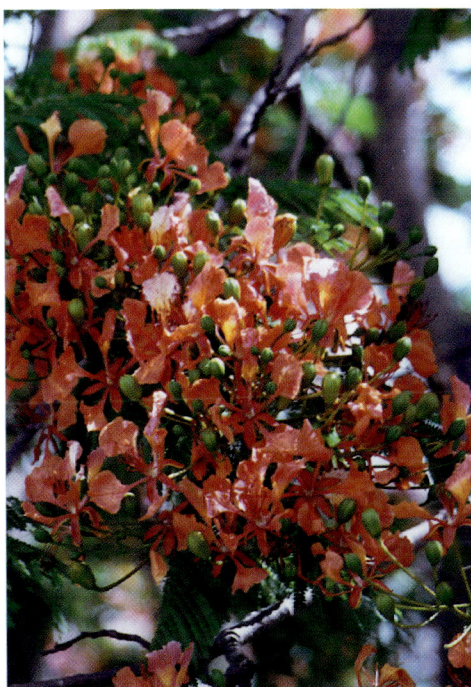

SILK-FLOSS TREE

(Chorisia speciosa)

FEATURES: Evergreen or semi-deciduous; height to 7–8 m; spread to 4–5 m. Non-invasive.

FLOWERS: Can be 10–15 cm across and are a vivid pink. The tree can shed all its leaves briefly just before flowering, making it as spectacular as a jacaranda, which has a similar habit. The flowers, inspected up close, are speckled.

LEAVES: Form a broad span of five or more leaves in a circle, soft green, as are many of the sub-tropical species. Not much shade.

WHERE: Very rare in cultivation in Perth. It will grow from the coast through to the hills in ordinary soil. A rich soil will encourage a greater leafy canopy if shade is required.

USES: Great to mix in with a tropical-look garden. The flowers are a sight to behold. Grow mainly as a specimen.

COMMENTS: Used extensively in the streets surrounding Disneyland in California. The trunk can develop a mass of thorns which, while intriguing, can hurt. They can be sheared off. The trunk and branches remain green and provide a perfect foil for the profuse flowers that cover the tree. There is a good specimen on Welshpool Road in Lesmurdie. The name comes from the silky threads that are produced with the seed.

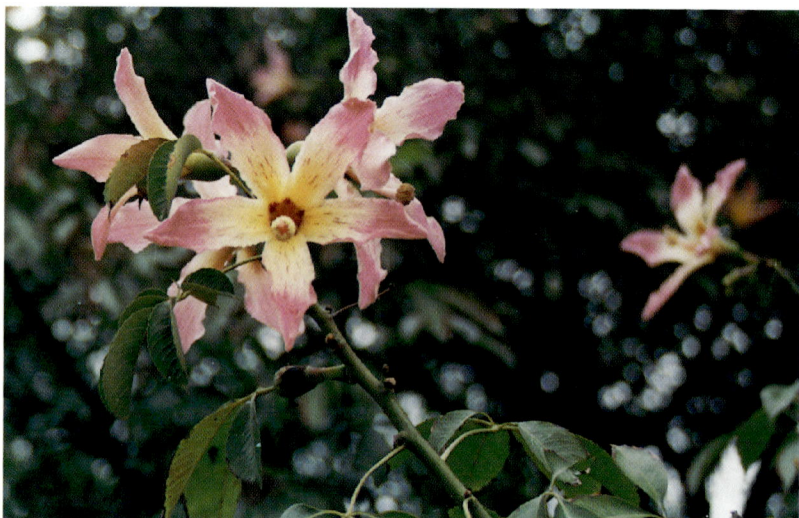

SILVER BIRCH

(*Betula pendula*)

FEATURES: Deciduous; height to 6–8 m; spread 1–2 m. Non-invasive.

FLOWERS: Elongated catkins 1–2 mm wide. Insignificant otherwise.

LEAVES: Shield-shaped, very attractive, hanging downwards. Casts medium shade. Leaves turn a matt butter colour in autumn which makes them very pretty.

WHERE: Fine by the coast in a protected courtyard or away from strong winds. Best planted in groups of three or more in varying sizes. The very small root system makes them ideal in tight corners or even in large containers. Water well over summer and mulch deeply. They look great in the hills where there is room for a grove of birches. Fast growing.

COMMENTS: Nothing is quite as reminiscent of northern Europe as the Silver Birch. There are many varieties around the world but for Perth the Silver Birch is the best. The almost white trunk stands out well with the deep green leaves. They work in well with a cottage garden. There are some nice specimens in a neglected garden bed at Royal Perth Hospital entrance on Wellington Street.

CULTIVARS: Weeping Birch (*Betula pendula* 'Youngii') is a grafted specimen. Used as a stand-alone feature, it can be planted with a small group of normal Silver Birches as a background.

Cut-leaf Birch (*Betula pendula* 'Dalecarlica') has a finely dissected leaf. Best grown in a very sheltered area away from very strong winds. A superb specimen if the conditions are right.

SIMON'S POPLAR

(Populus simonii)

FEATURES: Deciduous; height to 8 m; spread to 2–3 m. Non-invasive.

FLOWERS: Inconspicuous.

LEAVES: Elliptic, dark green, very attractive, turning a soft butter colour prior to leaf drop.

WHERE: This poplar has a slight spread at its top but is basically an upright tree. Very useful in tight areas with narrow garden beds. It is a very small poplar with the distinct advantage of not suckering. It does well from the coast to the hills. It enjoys a good watering; but is not fussy about the soil.

USES: Good as a narrow screen in a long line. Will provide good shade without being overpowering. Quite a handsome tree if planted on its own.

COMMENTS: The trunk is a soft silvery green, patterned by the old leaf scars. Its ability to adapt makes it very useful.

VARIETIES: The Cottonwood (*Populus deltoides*) is a broadly spreading poplar with large leaves. It, too, has a silvery trunk, but grows to large proportions. There are some in the Memorial Rose Garden on Stirling Highway, Nedlands.

The Aspen (*Populus tremuloides*) is a smaller poplar with the interesting habit of having leaves that shake in the wind, hence its common name, 'Trembling Aspen'.

SMOKEBUSH

(*Cotinus coggygria*) also known as Smoke Tree

FEATURES: Deciduous; height to 3–4 m; spread to 2 m. Non-invasive.

FLOWERS: Inconspicuous tiny flowers on the end of a panicle. When the seed sets they produce very fine filaments which look for all the world like smoke.

LEAVES: Oval and a milky-green. The green-leaf form is nothing special until autumn when it looks as though it is on fire. Splendid reds and oranges decorate the tree.

WHERE: Seems to grow quite happily in the middle suburbs of Perth and the hills. Likes a rich, well-drained soil with cool, moist roots.

USES: As an ornamental. It will develop a dense covering of leaves.

COMMENTS: I couldn't believe the first Smokebush I saw in full colour in autumn. I had to keep going back over and over to take in the colours. A gorgeous small tree.

CULTIVARS: 'Purpurea', a form with deep milky purple leaves. Quite an accent in the garden as the colour is rare; good autumn colour, too.

'Royal Flame' has a purple leaf with the edge tinged a fiery red. Otherwise same as 'Purpurea'.

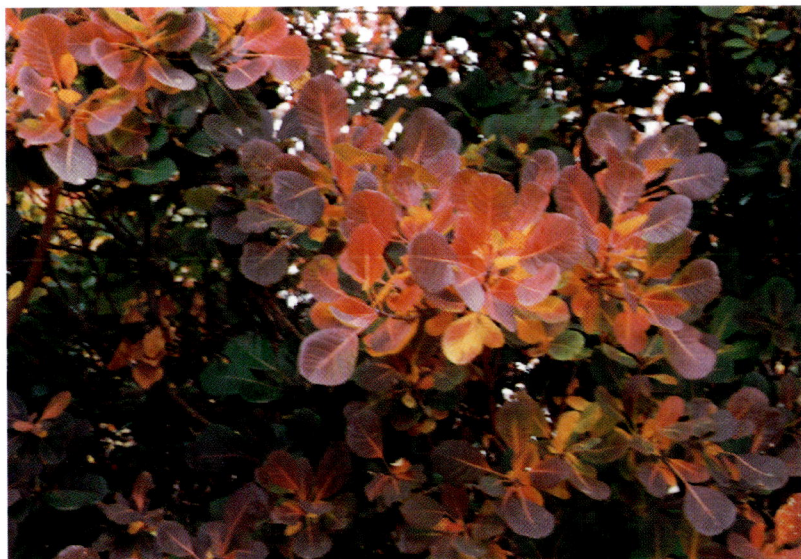

SOUTH AFRICAN TULIP TREE

(Spathodea campanulata)

FEATURES: Evergreen to semi-deciduous; height to 10 m; spread to 7–8 m. Non-invasive.

FLOWERS: A rich, bright orange, tulip-shaped flower that stands upright on the tree. The flowers are produced in a bunch that opens one or two at a time and does so for many months. Flowering is late summer to autumn.

LEAVES: Pinnate, thick, dark green and hairy, as is the trunk when young. It gives a dense shade and looks cool and lush in the garden.

WHERE: Not at all fussy except a dislike for frost-prone hills areas. Otherwise will grow from the coast to the hills. Likes a rich soil and good summer watering.

USES: A spectacular flowering and shade tree. Often produces multi-trunks which will allow for a shorter tree. These can be pruned off to leave a single trunk and if necessary the top can be removed to shorten the tree and encourage branching.

COMMENTS: The tropical and subtropical trees certainly seem to put on the boldest show of flowers, and this tree is no exception. Quite a few are growing around Perth — there is a small specimen hanging over a fence in Selby Street, Wembley, and a fine tree in South Terrace, Kensington, another in Gilbertson Road, Kardinya. Later to flower than many of its counterparts, it adds a flash of colour when the skies are starting to cloud over.

SWAMP CYPRESS

(Taxodium distichum)

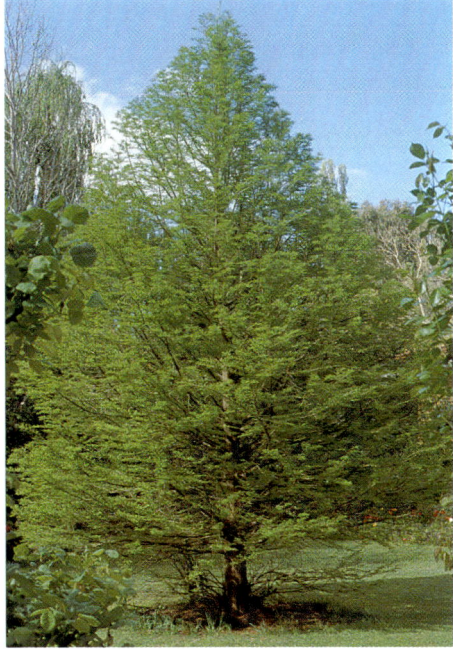

FEATURES: Deciduous; height to 10 m; spread to 5 m. Non-invasive.

FLOWERS: Insignificant.

LEAVES: Very small, ferny and a little like jacaranda leaflets. Not at all like a conifer, but that's what it is, and a deciduous one at that! In autumn the leaves turn a rich, rusty-brown before falling. Not everyone's cup of tea but I love it. From spring onwards the leaves are a soft green.

WHERE: Well, if you have a swamp in your backyard or a very wet area, this is the tree for you. In their natural habitat (south-eastern United States), they grow in a metre of water. In order to obtain oxygen they develop 'knees' which will grow above the water, and it is through these pneumatophores that they breathe. Strangely enough, they are equally at home in just plain old regular soil. Anywhere in Perth.

USES: This tree produces an extremely light, beautiful timber, prized by woodworkers. Its water resistance is remarkable! As a specimen tree it is very shapely, with a conical habit.

COMMENTS: Next time you are driving past the old brewery towards Perth, pull into that little park just after it. You will have to get out of the car to have a look at the lovely specimen growing there – it is a very happy Swamp Cypress. There are four large specimens along a driveway just after the Reid Highway on the right on Middle Swan Road. This amazing tree will keep guests talking for hours at a dinner-party!

TABEBUIA

(Tabebuia rosea)

FEATURES: Evergreen to semi-deciduous; height to 8–10 m; spread to 5–8 m. Non-invasive.

FLOWERS: Hot pink, trumpet-like, completely covering the tree. The tabebuia may shed all its leaves just before flowering, and if it does the effect is stunning. The area under the tree will also be a hot pink carpet when they fall.

LEAVES: Hairy, a brownish green, almost khaki colour, arranged in groups at the end of a stem. Good dappled shade.

WHERE: Quite happy on the coastal plain, but try to avoid howling winds, and watch the frosts in the hills. Build up the soil with rotted manure and water well over summer.

USES: If you want to meet a lot of people grow one of these in your front yard. They will queue up when it is in flower, asking what it is. Definitely an ornamental, giving good shade.

COMMENTS: A year or so ago, one of the gardening magazines featured this tree in full flower on a double-page spread. We sold every tree we had in about five hours! Not at all well known in Perth, it thrives here. Worth demolishing your house to plant one!

SIMILAR SPECIES: Silver Trumpet Tree (*Tabebuia argentea*). This tree has silvery-green leaves with yellow trumpet flowers arranged in balls. Very showy.

TIPUANA

(*Tipuana tipu*) also known as Pride of Bolivia

FEATURES: Semi-deciduous; height to 10–12 m; spread to 10 m. Invasive.

FLOWERS: Small, orange and ruffled. Carried all over the tree in early summer.

LEAVES: Small, paired, offering some shade, but not dense.

WHERE: Fine from the coast to the hills. Tolerates poor soil, but does better in a rich soil with good watering over summer.

USES: This is a spreading tree that can be kept manageable by heavy pruning every few years. Excellent for a tropical look in the garden. Casts good shade if well looked after but it is hard to grow anything underneath because of the roots. If you do plant one, hand water it well for a couple of summers to encourage a deeper root system.

COMMENTS: Now being used extensively as street trees around Perth. There are some good examples in the gardens outside Scitech Discovery Centre, and some very mature specimens in Harold Boas Gardens in West Perth.

HANDY HINT

BULBS can be lifted when all the leaves have withered and dried. If dug carefully you may find a 'set'. This is a new bulb produced by the parent plant. Sets can be separated and planted next season, and whilst they will grow into a mature plant they may take a few years to reach flowering. Dug bulbs are best stored in old onion bags or an old pair of pantihose, if dad has finished with them, and hung in a cool, draught-free area like a laundry until next season.

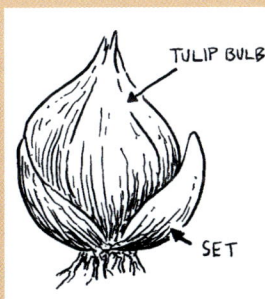

TULIP BULB

SET

TOONA

(*Cedrela sinensis*) also known as Chinese Cedar

FEATURES: Deciduous; height to 6–7 m; spread to 1–1.5 m. Non-invasive.

FLOWERS: Inconspicuous.

LEAVES: Like a fern leaf, pointed with indentations. Hot pink in spring when new growth appears. The leaves smell of onion if crushed, and have been used in cooking. The summer colour is a mottled green with white flashes. Autumn colour is old gold.

WHERE: Anywhere you like. They thrive in Perth, except on the ocean front. Soil type is of little concern, in fact they do well in degraded soil. A richer soil will make them look better, though.

USES: Very good in narrow or confined areas. It is not a wide tree and a tight corner would be fine. Very decorative in spring.

COMMENTS: I call this the 'accident' tree, because there are three growing just over a crest in Glen Forrest on Hardy Road. People new to the area come over the crest, spot these vivid pink trees and stop looking at the road — whammo! I suppose it really is unusual to see a tree covered in hot pink leaves.

SIMILAR SPECIES: *Toona australis*. Doesn't have pink leaves but a lovely orangey flower. This is native to the Blue Mountains of NSW. There is extensive use of the highly prized timber in the new Parliament House, Canberra. Bit of a waste if you ask me.

TUPELO

(*Nyssa sylvatica*) also known as Pepperidge

FEATURES: Deciduous; height to 5 m; spread to 2 m. Non-invasive.

FLOWERS: Insignificant.

LEAVES: Long, oval, mid-green, dainty. Autumn colour is orange through to scarlet.

WHERE: Grows naturally by American watercourses, and is named after a Greek god of water, Nyssa. This tree loves water but is just as at home in a regular garden. Best in a protected area away from violent wind. Middle suburbs to the hills.

USES: Purely as a small, ornamental, attractive tree.

COMMENTS: One of the loveliest of small trees. We grow it for the autumnal hues. Water it well over the hot months and it will reward you.

HANDY HINT

WARNING. Don't import illegally! When on an overseas trip, it is very tempting to pop a few seeds from an attractive tree or shrub into your pocket. The rationale is that they look clean, and a few little seeds couldn't possibly hurt anything. Nothing could be further from the truth. We are privileged in Western Australia by being isolated from all other States and countries by desert, distance and ocean. All the exotic diseases we have here have been imported by people. This number has been increasing, and we now have to spend millions to combat them. Perth is one of the very few western capitals in the world that doesn't have Dutch Elm Disease, and in a few decades we may have a thriving business of exporting elms back around the world. But not if someone brings in the disease! The penalty alone for being caught bypassing quarantine is substantial. There are avenues for importing seed, and I am sure just about anything anyone could want can be found within Australia — legally.

VARIEGATED PITTOSPORUM

(*Pittosporum eugenioides* 'Variegatum')

FEATURES: Evergreen; height to 6–7 m; spread to 3–4 m. Non-invasive.

FLOWERS: Insignificant.

LEAVES: Have a variegation that is not pronounced but produces a wonderful effect – a soft green, similar to the colour of avocado flesh. A very striking tree, densely foliated with this glowing colour.

WHERE: This pittosporum makes a marvellous hedge or screen. The stems are almost black and tend to highlight the leaves. Not good close to the coast, but fine in the middle suburbs and hills. They like a rich soil and good watering over summer.

USES: Because they are so dense they will block out anything behind them. Planted along a back fence is a good idea as the eye is drawn to the tree rather than the fence or neighbouring property.

COMMENTS: I first saw this tree used extensively in Victoria, from the north right into the Dandenongs. I never got bored by the soft green colour and have been sold on them ever since.

CULTIVARS: *Pittosporum tenuifolium* 'Tom Thumb' is a dwarf tree with lustrous ruby leaves, very bushy. It makes an excellent accent plant.

Pittosporum tenuifolium 'Limelight' is well named. Its colour is a soft lime and it grows in a more columnar form. Quite a rare colour and well used for effect in any garden.

Pittosporum tenuifolium 'Irene Paterson' has very creamy new growth that matures to a green marked with white (see photo). A very beautiful form of pittosporum.

WEEPING APRICOT

(Prunus mume 'Pendula')

FEATURES: Deciduous; usually to 2 m in height with a 1–2 m spread. Non-invasive root system.

FLOWERS: Massed pink blossom, with prominent yellow stamens, cascading down the weeping branches. Flowers start at the top of the branches and eventually unfold down to ground level.

LEAVES: Typical of apricots, oval coming to a point.

WHERE: Apricots are very hardy once established; best planted in winter with the weeping branches cut back hard at planting. Full sun is a requirement and all soil types are suitable.

USES: As a specimen tree. Can be containerised and grown in a large tub, although it looks just as good in the ground.

COMMENTS: The long, weeping branches reach from the crown all the way to the ground, and when covered in the soft pink apricot blossom look spectacular.

HANDY HINT

POTTING. To get that old mossy look on fresh terracotta, mix yoghurt and chook poo, in equal proportions. Paint it on the outside of the pot and wait a couple of weeks. Spore from moss is often airborne, and when alighting on your pot will find a perfect environment on which to grow. Before you know it, it will look older than me! By the way, the yoghurt doesn't need to be flavoured.

WEEPING MULBERRY

(Morus alba 'Pendula')

FEATURES: Deciduous; height to 3 m; spread to 2 m. Non-invasive.

FLOWERS: Insignificant.

LEAVES: Large, dark green, typical mulberry leaves. They turn a nice yellow in autumn.

WHERE: This very hardy tree can be grown anywhere from the coastal plain to the hills. It will take strong wind but may grow a little lop-sided. It loves a rich soil, good mulching and good watering over summer.

USES: The Weeping Mulberry has very long, water-falling branches that reach the ground and are covered in leaves. They look magnificent if you plant them where they can be walked around and enjoyed from a distance. The weeping nature means that they won't grow too tall. Ideally, should be planted as a specimen.

COMMENTS: Careful pruning, best explained by an experienced nurseryman, can make a Weeping Mulberry into a cool, outdoor, leafy arbour that you can sit under. It is possible to grow a 'door' into it through which to enter — kids love them. This tree produces small dry fruits that are marvellous in muffins. Very little problem with the juice staining carpets. Every garden in Perth should have one.

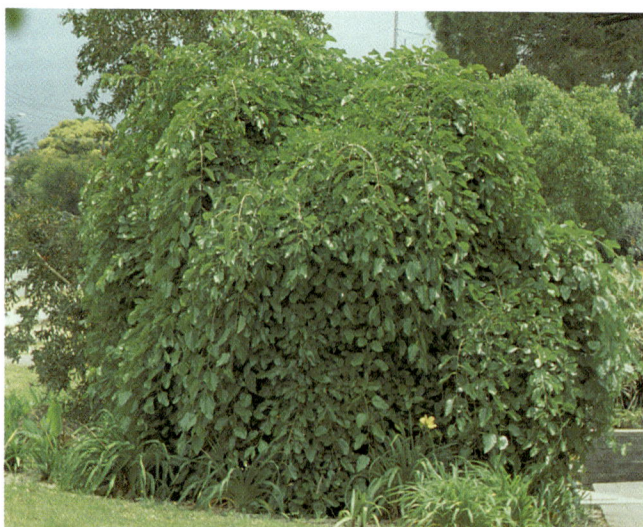

WEEPING SOPHORA

(*Sophora japonica* 'Pendula')

FEATURES: Deciduous; height to 2–2.5 m; spread to 1–1.5 m. Non-invasive.

FLOWERS: Pea-shaped, white, in a panicle. Although the tree form will do this every year, the Weeping Sophora is sporadic in flowering, sometimes with years between flowering.

LEAVES: Dark green, oval, the size of a ten cent coin. On the weeping form they are carried on branches that hang downwards. Butter-coloured in autumn.

WHERE: Anywhere not subject to strong winds. A rich soil is required, with good summer watering. Fine from the coast to the hills.

USES: Only as a very special ornamental small tree. Will grow happily in a large pot.

COMMENTS: The Weeping Sophora is one of the loveliest of all the small specimen trees. A little pruning to shape can bring the branches cascading down to the ground. They look fabulous grown over a pond or running water.

VARIETIES: The tree form of the sophora, onto which the weeper is grafted, is a very beautiful tree to 8 m. It will flower each spring with a mass of pure white flowers. The trunk is very interesting too, very dark green with grey patches.

S H R U B S

ALBERTA MAGNA

(Alberta magna)

FEATURES: Evergreen; height to 2 m; spread to 0.5 m Non-invasive.

FLOWERS: Tubular, with a flared tip, orange-red, usually in late summer.

LEAVES: Large and oval to elongated, glossy dark green. They give the plant an overall lustrous look, as they are often as large as a hand.

WHERE: A native of South Africa, it is happy from the hills to the coast, although protection from frost is advised. Give it good watering while establishing the plant and deep mulch as it prefers a rich soil.

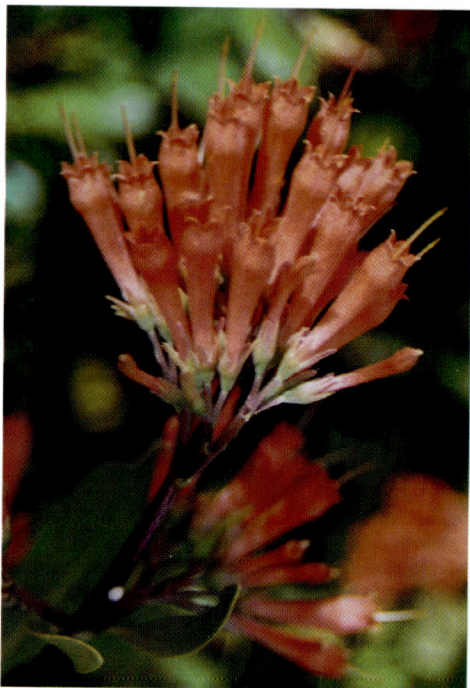

USES: Alberta magna has a great application as a hedging shrub. It can be clipped and shaped to make an informal screen or windbreak. It can be used to good effect as a background plant to show off perhaps the variegated foliage of a shrub planted in front of it.

COMMENTS: Not well known, this is an attractive shrub that should be grown more widely than it is. The flowers are quite pretty and stand out well. Ideal for screening or privacy.

HANDY HINT

LIGHT. Almost all plants are phototropic, which means they grow towards light that they need to photosynthesise. When growing potted plants (topiary in particular) it is a good idea to turn them around from time to time so that the reverse side is pointing towards the lightest area. You may find some dying back on the darker side of a specimen or it may be developing a lean — both of which effects can be caused by lack of light. A twist every few months will keep it well balanced.

ALTERNANTHERA

(Alternanthera)

FEATURES: Evergreen; height varies with cultivar, usually to 30 cm. Spread to 50 cm. Non-invasive.

FLOWERS: Unusual inasmuch as they are creamy little balls or tufts. Insignificant.

LEAVES: Alternantheras are grown for the many different colour combinations they have to offer, from the deep blood-red of A. *dentata* to the green-apricot of 'Apricot Splash'. The leaves can be thin and strap-like, or broad and oval.

WHERE: These low shrubs are grown mostly on the coastal plain, although they will do well in a frost-free hills environment. They enjoy good watering over summer and the odd application of an all-purpose fertiliser.

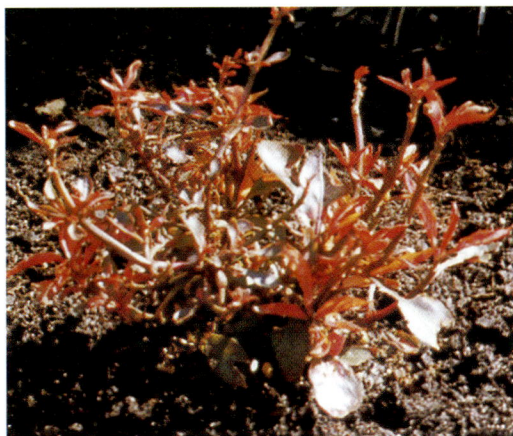

USES: Often used as a border or edging because they are so low-growing and their colours make for a very definite demarcation. I have seen the name of a golf-course spelled out with alternanthera, and very effective it was.

COMMENTS: When you see this plant used in long rows, straight or wavy, and another coloured cultivar laid next to it in another row, the effect is striking. A very effective and easy-to-grow accent plant. There must be about ten or so different colours and variations to choose from. Although they may not need it, they can be clipped to tighten the bush up a little without harm.

BARBERRY

(Berberis)

FEATURES:
Deciduous; height to
1–1.2 m; spread to
1 m. Non-invasive.

FLOWERS:
Inconspicuous.

LEAVES: Small and richly coloured, the berberis offers some of the most beautiful foliage available. *B. atropurpureum* has deep purple in small, almost box-like leaves, whilst the cultivar 'Rose Glow' has oval leaves, the new growth pink blotched with white and the older growth a deep pink-red. 'Golden Ring' has a bright yellow new flush of growth, deepening to a rich gold.

WHERE: Berberis will grow from the coast through to the hills. Try to avoid howling coastal winds as they will grow with the wind, and be bald on the windward side. These shrubs can be drought-tolerant once established and don't seem to mind fairly average soil. I find that a good slop of water over summer and some all-purpose fertiliser moves them along quite quickly.

USES: Although a single specimen is a marvellous accent in the garden, or maybe two or three cultivars near each other, I still prefer this shrub for hedging. The more you clip it into shape, the more you encourage that vivid flush of new, coloured growth.

COMMENTS: Berberis are prickly little devils, so wear some long gloves when clipping. The thorns will persuade a potential burglar to try elsewhere if he has to negotiate a hedge.

BOX

(Buxus sempervirens)

FEATURES: Evergreen; height to 2 m; spread to 1 m. Non-invasive.

FLOWERS: Insignificant, usually hidden within the foliage.

LEAVES: Small, rounded and a very dark green, about the size of a little finger nail. There are several species which have slightly different foliage, such as Dutch, Korean and Japanese Box. The most distinct species is *B. latifolia* which has elongated leaves and an upright growth habit.

WHERE: Box needs full sun to look its best, although it will take partial shade. You can grow it from the coast through to the hills. If you are subjected to salt spray, it is best to hose the plant down every now and then to wash it off. The secret of growing box is threefold: first plenty of sun, secondly good watering, and thirdly a regular supply of chook poo (if you live by the coast substitute cow poo).

USES: Box is grown mainly for a hedge, edge or topiary. Because it grows steadily, but not fast, it lends itself admirably to these uses. Commonly seen in formal gardens, box is the dark green, very neatly clipped miniature hedging that looks so good.

COMMENTS: English Box, as it is often called, will grow into a tree, although you are not likely to ever see one because it is kept clipped and shaped. For many hundreds of years it has been the favourite for topiary — carving spirals, balls, domes and animals out of its small leaves.

BROOM

(Cytisus)

FEATURES: Evergreen; height to 2 m; spread 1 m. Non-invasive.

FLOWERS: Pea flowers that come in a wide range of colours: white, yellow, brown, orange-yellow, peach, pink and a deep red. Late spring is when these shrubs do their thing. The branches are absolutely covered in blossom.

LEAVES: Modified to allow the plant to live in harsh circumstances, not the least of which is by the seaside. They are thin and scale-like making the shrub look scraggly when not in flower.

WHERE: These versatile shrubs will thrive in poor soil, and tough, almost drought, conditions. So you can grow them anywhere, from the coast to the hills. An occasional feed and water is all they need.

USES: If you have tough conditions, consider the cytisus, and even if you haven't, it is worth considering! For a very low-maintenance garden, this shrub is ideal. Grown only as an ornamental.

COMMENTS: It is hard to understand how such unprepossessing shrubs can look good. I often look at ours in summer or winter and wonder why we have them, only to be knocked out by the blossom in spring! Then I swear it is one of the most wonderful sights I have seen, only to repeat myself the following year.

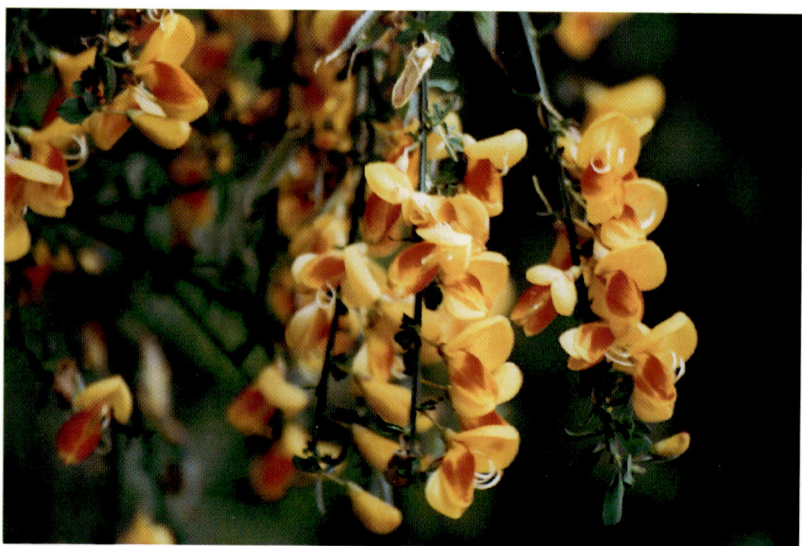

BRUNFELSIA

(*Brunfelsia latifolia*)

FEATURES: Evergreen; height to 1–1.5 m; spread to 1 m. Non-invasive.

FLOWERS: Quite unusual. They start off violet, turn mauve and then white – isn't that clever? The flower itself is about the size of a twenty cent coin and opens with five lobes. Brunfelsias are summer flowering and sweetly perfumed.

LEAVES: Nothing special, but a nice backdrop for the flowers.

WHERE: Brunfelsia is a warmth-loving shrub, both summer and winter, so find a nice hot spot in the garden. It will grow from the coast to the hills, but if there is a danger of frost plant it near the house for radiant heat in winter. Compost and old poo will keep it happy along with a good watering over the hot months.

USES: Best grown as an ornamental, it can make an effective hedge, especially when in flower.

COMMENTS: There is a lovely, very old shrub near the butterfly house in the Perth Zoo, well worth looking up in early summer. Sometimes called Yesterday, Today and Tomorrow because of its flowering habit, this lovely shrub is easy to grow.

HANDY HINT

HEAVY SOIL. When planting in heavy clay soil, cut the mix you put back in the hole with 50% original clay and 50% free-draining mix. If you use only a free-draining mix you will create a swimming pool and drown your tree or shrub. There are very few trees or shrubs that will last beyond four days of waterlogging.

BUDDLEIA

(*Buddleja*)

FEATURES: Evergreen or deciduous; height to 1–2 m; spread 1 m. Non-invasive.

FLOWERS: In *B. davidii* range in colour from pure white, pink, cerise and gold to crimson and scarlet. The flower sprays are cone-shaped and made up of hundreds of tiny florets. *B. latifolia* has massed branches, covered in long strands of pink-mauve flowers, that will weep to the ground. The flowering period is summer and autumn.

LEAVES: An elongated oval in *B. davidii*, with a silvery underside. They are very attractive in a gentle breeze as they flash silver. The other varieties have long, dark green leaves.

WHERE: Fine from near the coast to the hills. They are cold-tolerant but beware of heavy frosts. Buddleias like a rich soil, plenty of rotted poo and good watering over the hot months. As they can get tall and leggy it is best to cut them back after flowering.

USES: As an ornamental. These shrubs are very pretty when in full flower, and two or three different coloured varieties close to each other give a wonderful effect.

COMMENTS: Not only do they have a spicy fragrance but the leaves are aromatic, too! Perhaps this is why they are sometimes called Butterfly Bush; they will certainly attract butterflies to your garden. It is worth a visit to Bickley Valley Herbs and ask Kareena to show you her buddleia. I won't say any more.

CAT'S WHISKERS

(Orthosiphon aristatus)

FEATURES: Evergreen; height to 75 cm; spread to 50 cm. Non-invasive.

FLOWERS: A small cone of compressed flowers that emerges from the terminal growth and slowly swell until they are about 70 mm high. They then open in a whorl, starting from the base, and thrust out very long anthers that look just like a cat's whiskers. This continues all the way up until the whole head is open. Mostly white or a soft purple, these are very beautiful flowers that may continue for six months.

LEAVES: Rather ordinary for such spectacular flowers. Thick, oval with furrows and paired up the stem.

WHERE: Orthosiphon is a native of northern Australia and enjoys a warm to hot part of the garden, rich soil and good watering when it's hot. Boost them with an all-purpose fertiliser in spring and again at Christmas. No good in a frost-affected area, otherwise fine from the coast to the hills.

USES: A delightful addition to any garden. The best effect is in massed planting. They would look quite at home in a cottage garden. They are used to great advantage as a loose hedge, when after flowering, they can be cut back hard to shoot away again the following spring.

COMMENTS: Although they are easily grown, one rarely sees this gorgeous flower in Perth gardens. They can be found, however, and are well worth the hunt. If you live in a confined area, grow them in pots, they will perform just as well.

CAMELLIA

(Camellia)

FEATURES: Evergreen; height to 5 m; spread to 3 m. Non-invasive.

FLOWERS: Years ago I was so taken with the beauty of camellias that I resolved to collect one of each. Later I found out that there are more than 30,000 to choose from — blimey! It is difficult to describe such a vast array of glorious blooms; suffice it to say that it is hard to go wrong with any of them. The *sasanquas* generally have smaller blooms and are the sun-hardy camellias, while the *japonicas* are much larger, many with double and peony-shaped flowers, and still quite tolerant of the sun. The *reticulatas* have massive double blossoms that are breathtaking and need to be protected from the full blast of the sun. *Williamsii* are a smaller group of hybrids that are sun-tolerant. The colour range starts with pure white and then seems to visit all shades of the red range, to the deepest reds and scarlets. There is even a yellow camellia (*Camellia chrysantha*).

LEAVES: A glossy dark green, small in the *sasanquas* and larger in the *japonicas*. The occasional leaf may throw a cream streak which is quite normal. The darkness of the leaves helps heighten the richness of the flowers.

WHERE: Camellias prefer an acidic soil, a cool root run, good watering and mulching. They do very well in the hills and on the coastal plain, but the coast itself is generally too alkaline. The soil can be acidified or the camellias grown in containers if too much limestone is present. Bayswater has a memorial camellia garden on Guildford Road near King Edward Street with a good selection growing in full sun, well worth a visit. *Sasanquas* usually flower earlier than *japonicas*.

USES: Camellias come in many sizes, from miniatures through to the very large shrubs, so there is one to suit all gardens. Against a fence is a favourite spot, or to hide an unsightly building. Sasanquas are coming into their own as a hedging shrub.

COMMENTS: One camellia worth mentioning is the lovely weeping 'Mandy'. It has drooping branches that carry petite pink blossoms all the way to the tip. Gently perfumed and the softest of pinks, it looks wonderful in a container. With careful selection you can have camellias flowering from March to November.

HANDY HINT

STAKING. I rarely stake trees, preferring to force them to develop a deep root system to counteract wind movement. Trees that depend on a stake can be blown over once the stake is removed. If you feel it is necessary to stake a tree it is better to use three stakes placed in a triangle about 75 cm from the trunk, and then secure the tree to all three using old pantihose — just make sure you're not still wearing them first.

TRIANGULAR STAKE SYSTEM

←75cm

CEANOTHUS 'BLUE PACIFIC'

(Ceanothus impressus)

FEATURES: Evergreen; height to 1–1.2 m; spread to 1 m. Non-invasive.

FLOWERS: Pretty, deep blue puff-balls carried all over the canopy of the shrub. Flowering commonly starts in late winter to early spring and adds a rich welcome colour to herald in spring.

LEAVES: Thickish, leathery and rather nicely shield-shaped. The green is a light colour and offsets the flowers beautifully.

WHERE: One of the great advantages of this shrub is its ability to grow right by the coast. The winds and the salt spray don't seem to bother it much, with the odd hose down to wash off salt build-up, they thrive. You can grow them all the way into the hills without a problem. They will take harsh conditions, but good watering over summer and a decent mulch will make them perform well.

USES: Ceanothus make a good hedge or a border, the occasional clip to make a nice informal structure. On their own they look just as good as a garden accent. It is rare to find so lovely a blue on such a tough shrub.

COMMENTS: Native to the California region of the United States, ceanothus come from some pretty tough terrain, so no wonder they do well here. It's worth keeping an eye out for C. *impressus* 'Blue Cushion' which is a lower-growing form.

CESTRUM

(*Cestrum*)

FEATURES: Evergreen; height to 1–1.5 m; spread to 1 m. Non-invasive.

FLOWERS: C. *nocturnum* is a difficult shrub for nurserymen to sell. Although its sets quite respectable flowers, the heavenly perfume only comes out at night, long after the nursery is closed! In C. *nocturnum* the bunches of flowers are a whitish yellow. C. 'Newellii' has a mass of rich red tubular flowers with a soft pink at the flared end. C. *purpureum* has a similar collection of flowers that are a soft mauve, again with a pinkish end.

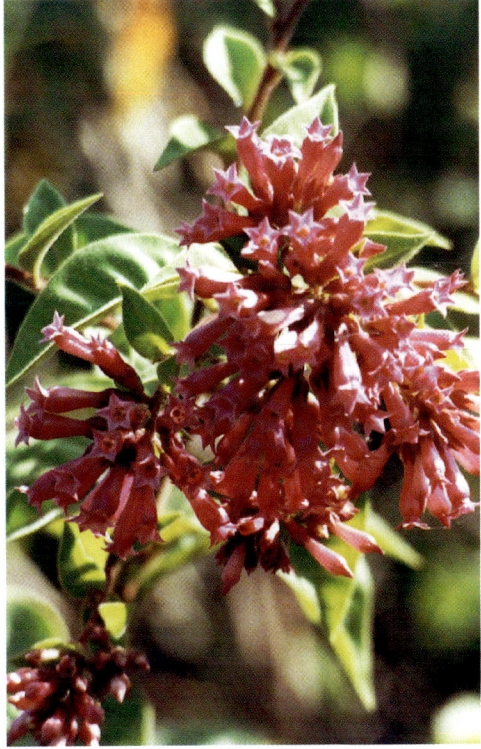

LEAVES: Tend to be long and narrow, a soft green except C. 'Newellii' which has much larger velvety leaves of a rich dark green.

WHERE: You can grow cestrum anywhere away from frost. They enjoy both winter and summer warmth but will take partial shade over summer. Try and avoid strong winds. Make sure that they have a good deep rich mulch and good watering over the hot months.

USES: Plant a C. *nocturnum* near an area where you entertain at night in summer to enjoy the fragrance. The other species will make a lovely accent in any garden and can be clipped to shape, after flowering, to keep them compact. Good for screening ugly sights like pencil pines.

COMMENTS: Some time ago we planted a C. *nocturnum* under our bedroom window and on those hot, still summer nights the smell of a rich apricot mousse floats through our window. Heaven on a stick!

CHASTE TREE

(*Vitex agnus-castus*)

FEATURES: Evergreen; height to 3 m; spread to 2 m. Non-invasive.

FLOWERS: Lavender-blue to mauve, borne on 10 cm trusses in summer. Each flower is tiny, but massed together they are very pretty.

LEAVES: Trifoliate, a long oval shape coming to a point. The new growth is purple, as is the underside of the mature leaf. There is a variegated form that has a soft green leaf covered in white blobs and makes a very attractive colour, especially from a short distance.

WHERE: Another tough customer. You can grow this wonderful shrub anywhere in the Perth region. They will tolerate a drought situation when established, thrive in poorish soil and don't mind being by the ocean. They will look lush if they receive a good drink every now and then and a handful of all-purpose fertiliser — but you certainly don't have to spoil them.

USES: They make an excellent soft hedge. The leaves are lovely to brush against, and the purple underside will flash its colour at you in a gentle breeze. The variegated form is also a handsome hedge: there is a good example at the bottom of Kalamunda Hill at the petrol station. Vitex can be pruned to give a tall dome shape, or be kept lower as a mound. Left on their own, they will reach a good size, but can always be trimmed back.

COMMENTS: The name comes from an ancient belief that these shrubs represent purity — nice, eh?

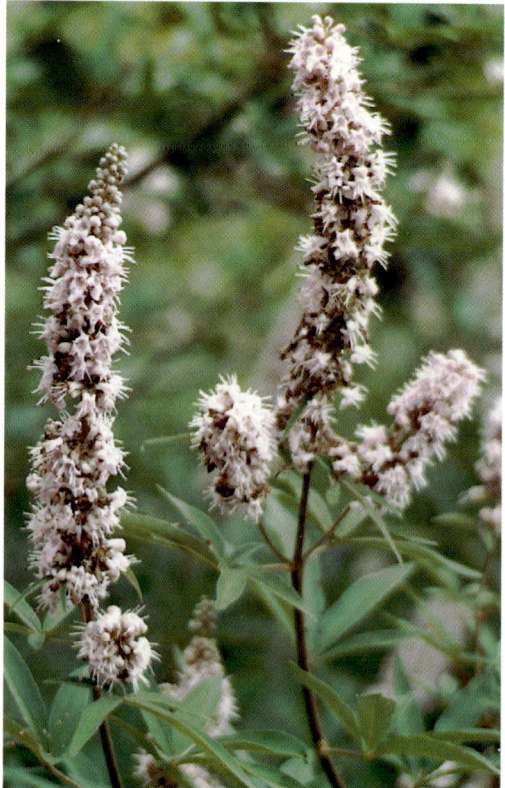

CHOISYA

(Choisya ternata)

FEATURES: Evergreen; height to 2 m; spread to 1 m. Non-invasive.

FLOWERS: Highly perfumed, not unlike orange blossom, white and star-shaped. They are massed all over the bush, from late spring to summer.

LEAVES: Dark green, rounded and come in threes, spreading outwards like a wheel. The mounded shrub is covered firstly by these leaves, and later by the blossom. There is a cultivar called 'Sundance' which has very attractive golden leaves.

WHERE: Not all that fussy about hot and cold, choisya don't like strong winds or an alkaline soil. Give them an acidic soil, plenty of well-rotted poo and a good watering over summer.

USES: Excellent for hiding things, like neighbours or ugly sheds. Choisya will make a good wide hedge, and if clipped will remain very bushy — best after flowering to encourage more flowers next season.

COMMENTS: When you smell the blossom of this shrub you could swear it's orange blossom, and for a good reason: they are close cousins. Just no oranges!

CONVOLVULUS CNEORUM

(Convolvulus cneorum)

FEATURES: Evergreen; height to 30–50 cm; spread to 50 cm. Non-invasive.

FLOWERS: White and large, the size of a fifty cent coin. The petals are joined to form a circle and the centre has yellow stamens. This shrub thrusts its flowers upwards so you don't have to bend over to enjoy them. Flowering starts in spring and is massed on the bush.

LEAVES: Silvery-green, elongated strap-shaped and rounded at the end. The silver appearance is due to many tiny hairs on the upper side of the leaf.

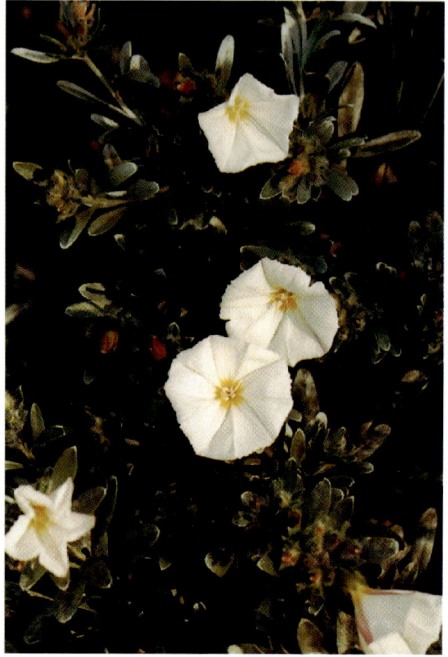

WHERE: A poor, sandy seaside soil is fine. Tolerant of salt spray and drought-resistant, this is an ideal shrub for the coast. Equally at home in all other areas, it also serves a good use in soil retention. Don't over-water it or be heavy-handed with fertiliser.

USES: This shrub fits in well with other seaside plants and the silvery foliage would work well with lavender and cistus. For areas requiring low maintenance, this is ideal and looks good in a mass planting.

HANDY HINT

BULBS. Bulbs such as tulips need a good feed with all-purpose fertiliser after flowering. This is when the bulb is making its bud for next season before going into summer dormancy. If you were to cut a tulip bulb in half, you could see the tiny bud right in the heart of the bulb. They like blood and bone too!

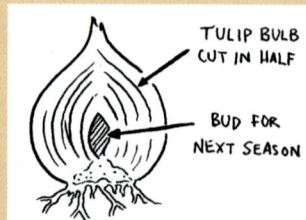

TULIP BULB CUT IN HALF

BUD FOR NEXT SEASON

COTONEASTER

(Cotoneaster)

FEATURES:
Evergreen/deciduous;
height to 1 m; spread to
1 m. Non-invasive.

FLOWERS: Four-petalled,
white, the size of a five
cent coin, followed by red
berries. Birds love these
berries and will be
attracted to the garden.
The flowers appear in
profusion all over the
shrub from spring to
summer; sometimes it is
hard to see the leaves.

LEAVES: Small and
rounded. The deciduous
forms can turn a nice
autumn colour, whilst the
evergreens will have the odd leaf turning crimson in autumn.

WHERE: These versatile shrubs will grow anywhere in Perth. Most
are small or prostrate and work well as a ground cover, or small hedge.
There is a good example at Botanic Golf on Burns Beach Road,
Wanneroo. Cotoneasters are quite undemanding in soil type and have
proven drought-resistance. Nonetheless, an all-purpose fertiliser in
spring and some summer watering will give you a good show.

USES: Good in a rockery, near cascading water or as a hedge. If
clipped they will bush up to a dense shrub, and can be formalised
quite successfully. As a ground cover they will spread nicely and cover
a good area. The commonly grown tree forms of cotoneaster are best
left to large country gardens.

COMMENTS: *Cotoneaster horizontalis* 'Variegatus' has variegated
leaves and makes an interesting accent in the garden. *C. dammeri* is
good for hedging, as is *C. microphylla*.

DAPHNE

(Daphne)

FEATURES: Evergreen; height to 1 m; spread to 40 cm. Non-invasive.

FLOWERS: White or pinkish, small, clustered at the end of a branch. Daphne is grown for its perfume — it is strong and slightly citrussy. Generally, when daphne is grown in the garden, it is the fragrance that lets you know it is flowering. It is one of the highlights in spring.

LEAVES: Thick and leathery, a dark green. They are oval, coming to a sharp point. Arranged just behind the flowers, they offer up the flowers to your olefactory glands.

WHERE: There are three major secrets to success with daphne. First, they must be virus-free stock when you buy them. Secondly, they need very good drainage — make sure a potting mix has plenty of coarse pine bark or similar. Thirdly, don't let them dry out; the roots need to be moist. Given that they prefer slightly acidic soil, you can grow them anywhere from the hills to the coast. They do well as potted specimens, taking a long time between repotting.

USES: They used to be popular in bridal bouquets or as a buttonhole for gentlemen. I like them in a pot so I can move them close to the house when they are flowering and enjoy their fragrance.

COMMENTS: Our climate generally isn't cold enough for most species, but as luck would have it, *D. odora*, which is the perfumed variety, does well here. There is a variegated form with a yellow margin in the leaf, very attractive.

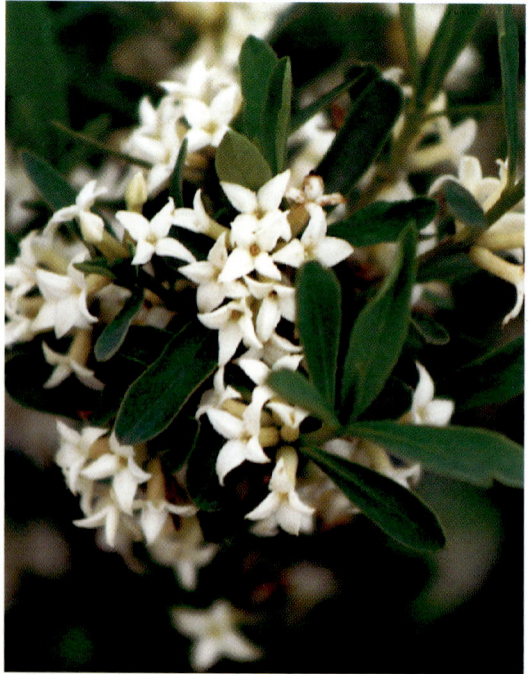

DAUBENTONIA

(Daubentonia tripetti [Sesbania grandiflora])

FEATURES: Evergreen; height to 2 m; spread to 1 m. Non-invasive.

FLOWERS: Hanging in wattles, pea-shaped like wisteria but a vivid scarlet. They normally start in mid-summer and cover the bush for a long time. The racemes are crowded with flowers and are around 20–30 cm in length.

LEAVES: Dark sea-green, pinnate with paired leaflets. They are not heavy on the bush, rather more dainty and work as a good foil for the flowers.

WHERE: There is not a problem growing daubentonia. I have grown them through frost and in a hot spot in summer without a problem. They will put on their summer show anywhere from the hills to the coast. Give them a rich soil and good summer watering.

USES: Very nice in a cottage garden as a backdrop to smaller shrubs, good on a fenceline, or just free-standing in a select part of your garden. The dappled but good light they let through makes them ideal in massed planting with a contrasting colour below.

COMMENTS: Sometimes called the Scarlet Wisteria Tree, it has a tree shape although quite small. The canopy spreads at the head of a tall trunk, giving an attractive shape. They rarely need pruning.

DEUTZIA

(Deutzia gracilis)

FEATURES: Deciduous; height to 1 m; spread to 0 .5 m. Non-invasive.

FLOWERS: Papery, flared, white bells or bonnets, carried in panicles and appearing in spring on terminal growth; then the shrub is a mass of colour and makes a dramatic show. *D. scabra* 'Candissima' is a double white form.

LEAVES: Long, dark green, sort of sardine-shaped, with a serrated edge. When not in flower, the shrub is quite handsome as an accent plant because of the leaves.

WHERE: Hailing from Japan, they do best in the hills or coastal plain, but not so well right by the coast unless they are well protected from the wind. They prefer an acidic soil that is rich in well-rotted poo and well mulched. Keep the water up to them over the hot months.

USES: Deutzias can be used to great effect in a cottage garden or in a more formal garden where they can be clipped to give a mounded effect. Nice used as an informal hedge.

COMMENTS: During summer you may find a water shoot growing up through the plant. It is best to leave this as it will carry a mass of blossom the following spring. Pruning can be carried out after flowering.

DICHORISANDRA

(Dichorisandra thyrsiflora)

FEATURES: Evergreen; height to 75 cm; spread to 50 cm. Non-invasive.

FLOWERS: In autumn you will notice what appears to be a small bluish bunch of tiny onions forming at the top of a stem. This slowly expands to reveal a very dark blue flower spike about 15 cm long. Once the whole spike is formed, each flower will open to the deepest blue with a very small collection of yellow stamens in the centre. The flowers last a very long time, and can be cut for indoors.

LEAVES: Arranged in a spiral that opens outwards. The leaves are fleshy and quite large. They are a very dark green and are carried on a thick, succulent stem. When out of flower they give a lush, tropical look to the garden.

WHERE: Avoid frost, although they are cold-tolerant. Best grown in a semi-shaded spot under the cover of shadecloth or a larger tree or shrub. These very beautiful shrubs are subtropical and enjoy humidity over summer, a soil rich in compost and good watering. They can be allowed to dry out a little over winter. They make an excellent potted specimen and the whole plant can be brought indoors when in flower.

USES: Use dichorisandra in a tropical-look garden as a foliage shrub with the added benefit of a spectacular flower. They will expand slowly into a clump, giving a dramatic effect.

COMMENTS: This Brazilian plant is often thought of as a ginger, but in fact it isn't, being related to the Wandering Jew. When I found my first plant I just couldn't believe the flower, and there is always great excitement when the next season's flowering begins.

DURANTA

(Duranta erecta [D. repens])

FEATURES: Evergreen; height to 2–3 m; spread to 2 m. Non-invasive.

FLOWERS: A soft blue or sometimes violet/blue, with four petals, two of which have a dark blue marking. The profusion of flowers is carried in drooping racemes, later producing beads of orange, poisonous berries. There is a smaller-growing, white-flowering form.

LEAVES: Oval and quite glossy, paired on a branch. The colour is a light green. *Duranta erecta* 'Sheena's Gold' is a cultivar that has rich golden foliage with new growth being a delicate translucent gold — most attractive. *D. erecta* 'Aussie Gold' looks as though it were designed for our olympic colours, green and gold in stripes along the leaf.

WHERE: Duranta has proven its ability to withstand harsh conditions. It makes a very good windbreak, grows in average soil and can go for long periods without a drink. It can be grown with much success from the hills to the coast.

USES: If it is watered regularly, the growth will be quite heavy and a better shrub will result. Best kept trained to a main trunk, unless they are being used as a windbreak when they can be kept in check by an annual pruning. Duranta makes a good large, tough hedge. The coloured-leaf forms are far more compact and can be used to good effect as accent plants.

COMMENTS: The benefit of durantas is the range of areas and conditions under which they can be grown. They are also very fast growing and lend themselves quite happily to clipping and shaping.

DWARF FLOWERING POMEGRANATE

(Punica granatum nana)

FEATURES: Semi-deciduous; height to 1 m; spread to 1 m. Non-invasive.

FLOWERS: Just like its larger fruiting cousin, but there the similarity ends. The dwarf form opens to reveal a mass of deep red ruffled petals — they always remind me of a flamenco dancer's skirt. Very pretty and lots of them over summer, the colour is vibrant and can be seen from a distance. Occasionally there is a flash of white in the flower.

LEAVES: Rounded and sparse. The colour is a softish green.

WHERE: Anywhere that has a well-drained soil, not overly rich, and where you can give them a drink. Once established they are very drought-tolerant and the odd bit of rain is sufficient. You can grow these shrubs anywhere you like in Perth, in full sun.

USES: The Dwarf Pomegranate is a little unruly in its habit, so park it in a bed with a bit of room. It can be clipped to shape easily and without harm. They look great in large terracotta containers by a pool where they receive a little more care and pruning to produce a ball shape.

COMMENTS: The ease of growing this shrub makes it a good one for a harsh garden that hasn't yet had time to be built up. The sandy soils of Perth suit it well, as does a coastal situation. The flowers are gorgeous.

ECHIUM

(Echium candicans [E. fastuosum])

FEATURES: Evergreen; height to 1–1.5 m; spread to 1 m. Non-invasive.

FLOWERS: The flowers are in a large cone filled with hundreds of tiny florets from the palest blue to a deep violet/blue. The buds begin to appear in late winter and open over spring, the whole cone growing to around 20 cm in length. They are on the tip of each branch and point upwards. There is also a red-flowering form.

LEAVES: Very long and thin, and sage green. The branches have very thick stems, as thick as a finger, with the leaves arranged in whorls around them.

WHERE: Echiums prefer a poor soil and moderate watering. They will grow anywhere in Perth and don't seem to be worried by wind. The silvery-grey foliage is a give-away in indicating a seaside plant. Echium will take all the coastal problems in its stride.

USES: Where little else will grow. The shrub will spread from the base and offer up long branches that eventually carry the flowers. They are best cut back hard after flowering to keep compact and encourage a flush of new flowers the following year. They make a bold statement in the garden and look very effective with other silvery-foliaged plants.

COMMENTS: Though distantly related to Paterson's Curse, don't throw you hands up in horror — this plant is easily controlled. The flowers are lightly fragrant and attract the birds and bees.

ESCALLONIA

(Escallonia bifida)

FEATURES: Evergreen; height to 1–1.5 m; spread to 1 m. Non-invasive.

FLOWERS: In spring on terminal shoots. The clusters of flowers are in the range of soft to dark pinks and are fragrant. The flowers themselves are small, bell-shaped and last quite a while.

LEAVES: Glossy and dark green, with a serrated edge. Very bushy plants have a heavy leaf growth.

WHERE: I have seen a lovely specimen in Augusta that defied all the odds, including terrible winds, and little watering, that looked just beautiful, clothed in a mass of pink flowers. They appear to be very hardy, enjoying a built-up soil and summer watering to establish. You can grow escallonia anywhere in Perth.

USES: This shrub is a good one to use to cover unsightly areas. Use it as a screen, windbreak or hedge. You can clip it to shape and encourage more flowering wood. It makes a nice single specimen and would be at home in a cottage garden.

COMMENTS: One of the old-fashioned shrubs, escallonias are being grown more and more for the ease of growing and the beauty of the flowers and foliage.

HANDY HINT

POTS. If you're stuck with a badly root bound area (usually the result of poor watering technique, see pages 13-14), plant what you want to grow in green plastic pots. If the pots are tiered, with the tallest at the back and the smallest at the front, you will have the effect of a lush garden bed. Just remember to watch the pots, as they tend to dry out quickly in the summer.

EUPHORBIA LEUCOCEPHALA

(Euphorbia leucocephala)

FEATURES: Deciduous; height to 1 m; spread to 0.5 m. Non-invasive.

FLOWERS: Like the poinsettia, *E. leucocephala* has insignificant flowers but a wonderful show of white modified bracts to draw your attention. This show is put on in winter when things in the garden can look a bit drab. The pure white of the bracts fairly glows. A cultivar, *E. leucocephala* 'Pink Finale', does the same, except that the final flush of colour turns from white to soft pink!

LEAVES: A very attractive small leaf, oblong in shape. During summer the colour is a soft green, with a matt finish to the leaf.

WHERE: The same as the poinsettias. Avoid frosts and grow in a warm protected part of the garden in a rich soil.

USES: These little beauties can be grown in pots and tubs for quite a while, the advantage being to move them to a place of prominence when they are in flower. They look lovely in the garden and are used extensively in Queensland for hedging where they look magnificent.

COMMENTS: Not widely grown in Perth, which is a surprise because they do well here. Well worth hunting them down; sometimes they go by the common name of Snowflake.

FLOWERING QUINCE, JAPONICA

(*Chaenomeles speciosa*)

FEATURES: Deciduous; height to 2 m; spread to 1 m. Non-invasive.

FLOWERS: About the size of a twenty cent coin when fully open. The flowers come in white, soft pink, light red, deep red, and my favourite, Apple Blossom. Appearing on the bare branches in spring before leaf set, the effect is brilliant, masses of glorious blossoms, sometimes followed by an edible fruit which is great for making quince jelly.

LEAVES: Attractive, rounded, sometimes turning good autumn colour before shedding.

WHERE: Coastal plain and hills in a rich, well-draining soil. Water well over summer until established. The Flowering Quince can be grown in large containers and trained up to a single trunk. This shrub will send up suckers close to the main trunk and can be lifted in winter and divided — separate one off and give it to granny for Christmas and watch her eyes light up!

USES: This delightful shrub will espalier really easily. You can train it on a trellis or wires to give it a formal look. An almost necessary addition to a cottage garden. Use the Flowering Quince as a backdrop for smaller shrubs.

COMMENTS: I am often surprised at how long the Flowering Quince keeps producing those beautiful blossoms, often well into summer. It can be very effective to plant two differently coloured varieties in the same hole for a startling effect. The spine-tipped stems make it a useful people-stopper.

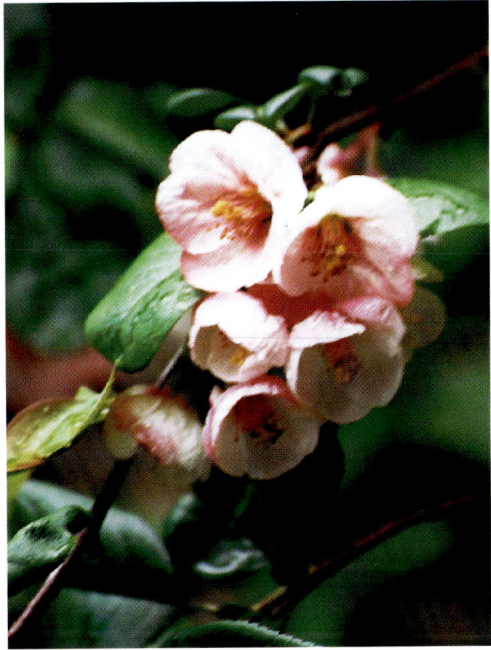

GARDENIA

(*Gardenia*)

FEATURES: Evergreen; height to 1.5 m; spread to 1 m. Non-invasive.

FLOWERS: White, highly perfumed, round with thick petals, from the size of a five cent coin in G. *augusta* to the size of an apple in G. *augusta* 'Professor Pucci', and many sizes in between. G. *augusta* 'Golden Magic' produces flowers that begin white, then turn to a rich cream and finally gold. G. *thunbergia* is a species with fragrant tubular flowers on a large bush, and G. 'Ocean Pearl' has small round flowers. They will flower over summer for ages!

LEAVES: Glossy, dark green, large and usually elongated. The foliage is most attractive, especially in mass plantings. G. *thunbergia* has strange, dark green, crinkled leaves that are long and pointed.

WHERE: Gardenias like an acidic soil, and although they are from the subtropics they take the cold fairly well. A frost will burn them, so beware. You can grow them from the hills to the coast, but in the latter area the soil must be acidified. They love a rich soil and plenty of water over the hot periods. Deep mulch and feed with an all-purpose fertiliser in spring, and again in summer.

USES: From a standard in a pot, to a hedge, to mass planting, to borders and edges to a single specimen — everyone needs at least one gardenia.

COMMENTS: I just love the way the flowers stand clear of that rich, dark green collection of leaves. On a standard, the mass of flowers is just at nose height — perfect! A mixture of the different varieties can give a lovely effect, for example, using perhaps G. *augusta* 'Florida' as a background and the miniature G. *augusta* 'Radicans' in the foreground.

GARRYA

(*Garrya elliptica* 'James Roof')

FEATURES: Evergreen; height to 3 m; spread to 1–2 m. Non-invasive.

FLOWERS: Greyish catkins about the thickness of a pencil hanging in very long tassels, up to 30 cm long. The shrub is covered with them in mid-winter.

LEAVES: Quite leathery and robust, not unlike a holly in shape but not as prickly, able to withstand quite strong wind and tough conditions. The topside of the leaf is green and the underside grey.

WHERE: Garrya is a native of California and grows happily in the hills overlooking Los Angeles. There the conditions are hot and dry over summer and quite wet in winter (sound familiar?). We can easily reproduce those conditions here and have a fine looking shrub.

USES: Really as a remarkable, winter-flowering shrub. You will be tempted to cut the flowers and use them in indoor decoration, which is fine. If you keep it compact with tip pruning, it will act as a good evergreen screen.

COMMENTS: If you haven't seen a garrya in flower before, you have a surprise in store! It is like something out of a sci-fi movie only quite lovely. The plant is dioecious which means there are separate male and female plants, and as with many things in nature it is the male that puts on the show. The cultivar 'James Roof' has extra long tassels. As they are now produced, with some difficulty, from cutting you can be assured of getting a male!

GOLDEN BELLS

(Forsythia suspensa)

FEATURES: Deciduous; height to 1.5–2 m; spread to 1 m. Non-invasive.

FLOWERS: Beautiful golden bells the size of a five cent coin, suspended on bare branches from winter to spring. The long, arching branches, weighed down with blossom, allow the flowers to hang downwards and the display is vibrant.

LEAVES: Densely arranged on the shrub and help tighten a hedge if this is what the shrub is used for.

WHERE: This is really one for the hills dwellers. Forsythia needs a cold and even frosty winter to set the flowers. If you can achieve this in a cold part of the garden, then definitely go for it — they are glorious. Plant into a rich soil, plenty of well-rotted cow poo and feed with an all-purpose fertiliser in early summer. Water well over summer.

USES: For a breathtaking hedge use forsythia. It will be one of the wonders of the world! It is usually grown as a single specimen plant.

COMMENTS: To keep the plant in check, eg. for a hedge, make sure that pruning is done only after flowering. *F. x intermedia* is a freer-flowering form of *F. suspensa*. I am always surprised by our forsythias. I go looking for the flowers and there is nothing to be seen, then one day when walking past there they are; they trick me every time.

GORDONIA

(Gordonia axillaris)

FEATURES: Evergreen; height to 2–3 m; spread to 1.5 m. Non-invasive.

FLOWERS: Up to 10 cm across, pure white, ruffled with a mass of prominent yellow stamens in the centre. The flowers are lightly perfumed and appear from late autumn through winter. Long-lasting as a cut flower, or float the flower itself in a bowl of water. The flowers will remain in good condition when they fall and make a beautiful display lying on the ground!

LEAVES: Very similar to a camellia leaf, to which they are closely related. They are a very dark green, serrated and the occasional leaf will turn brilliant red or gold in autumn.

WHERE: Gordonias, like camellias, prefer an acidic soil rich in well-rotted manures. Deep mulch the shrub and water well over hot months. If you can grow camellias then you can grow gordonias. All areas except the coastal limestone country, where they could be grown in a large container.

USES: The main reason is for the lustrous foliage and gorgeous flowers, but at the same time ugly areas can be well screened by a gordonia.

COMMENTS: Definitely one of my favourites and they grow very well here. I have seen a magnificent example growing in the mountains around Cairns, so they obviously like the tropics as well, even though they are very tolerant of cold.

HIBISCUS

(*Hibiscus*)

FEATURES: Evergreen or deciduous; height to 3 m; spread to 2–3 m. Non-invasive.

FLOWERS: *Hibiscus rosa-sinensis* is the large Hawaiian evergreen in profusion around Perth. These have been hybridised for many years and there are now hundreds from which to choose. The colours range from white to gold to scarlet and nearly every colour in between. *H. schizopetalus* is a Japanese species that has a filigree flower in a deep pink and hangs downwards. *H. syriacus* provides the deciduous varieties which offer a nice alternative to the Hawaiian forms. Often the flowers are doubles, ranging from pure white to mauve and an incredibly rich crimson. In general, *H. syriacus* flowers before new season's leaf set and makes a remarkable display.

LEAVES: Thick and leathery, large and profuse. The evergreens have quite large leaves, about the size of a hand, whereas the deciduous varieties have a smaller, more rounded leaf. Well cared for hibiscus will have deep green in the leaf and look lustrous.

WHERE: Hibiscus are notorius for being able to grow in rotten sand with little watering, and this is true ... but if you give them a rich soil, water them well and make sure they are warm, they will look fabulous. An all-purpose fertiliser application in spring and again in summer will keep them flowering for most of the year. They can be grown from the hills to the coast.

USES: To hide ugly buildings and neighbours, as a hedge, as a screen. A mixture of flowering colours is very effective. The deciduous varieties lend themselves beautifully to standardising.

COMMENTS: The Rose of Sharon, *H. mutabilis*, is a real winner. It flowers late, usually in autumn, and has pure white and deep pink flowers on the same shrub. These it carries high for all the world to see. There is nothing quite like hibiscus to keep the lovely little honeyeaters in your garden.

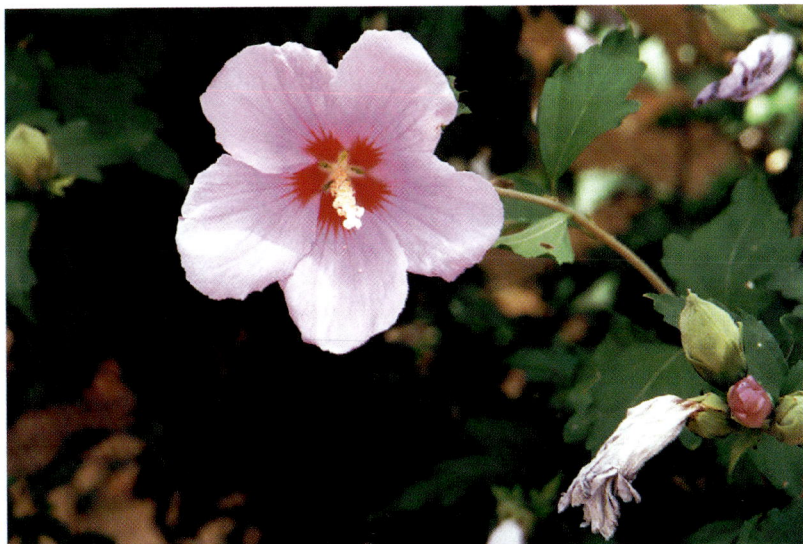

HANDY HINT

GRAFTED OR BUDDED TREES. Many trees are produced by grafting or budding because they cannot be grown from seed. Although a Claret Ash may produce seed, that seed won't produce a tree that turns claret red in autumn. The original Claret Ash was just one out of millions of Desert Ashes that showed a beneficial mutation — it turned the deepest red in autumn. It did this consistently every autumn, and finally buds taken from it were grafted onto normal Desert Ash. Come autumn, these, too, turned red because they were a clone of the original tree. The normal ash it was grafted onto is called the under-stock and the mutated form is called the mother stock. This is the reason a Claret Ash is more expensive than a Desert Ash. So, if you have been collecting seed from a budded or grafted tree, all you will have is the understock when it grows.

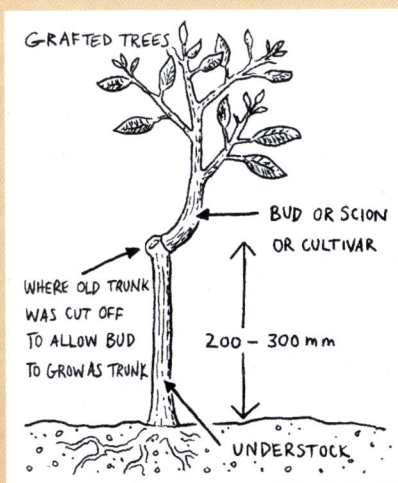

GRAFTED TREES

BUD OR SCION OR CULTIVAR

WHERE OLD TRUNK WAS CUT OFF TO ALLOW BUD TO GROW AS TRUNK

200 – 300 mm

UNDERSTOCK

HOLLY

(Ilex)

FEATURES: Evergreen; height to 3 m; spread to 2 m. Non-invasive.

FLOWERS: Quite small and white, followed by the red berry on the female shrub.

LEAVES: Very distinctive, commonly prickly and generally very dark green. These are the sprigs usually seen on Christmas cards. Some varieties produce very attractive leaf variegations. *Ilex altaclerensis* 'Golden King' has waves of creamy gold on a green leaf, while *I. aquifolium* 'Lawsonia' has green, cream and yellow all in one leaf. *Ilex aquifolium* 'Argentea Marginata' has a green leaf edged in cream, and *I. aquifolium* 'Golden Milkboy' shows a predominence of gold edged in green. *Ilex cornuta* has a large, solid green leaf.

WHERE: Hollies have always been thought to grow only in cold climates. They will, in fact, do well here. They love a moist, rich soil, good mulching and good watering until they are established. Grow them in the hills and on the coastal plain, and by the coast if the spot is protected.

USES: The Weeping Holly is a very beautiful specimen tree grafted at about two metres and allowed to weep out and down. Being evergreen, hollies make an excellent hedge; best to use *I. cornuta* which is the most popular hedging plant in Japan. The red berries are ripe in July, just in time for a Christmas-in-July celebration!

COMMENTS: The hollies have been used for hedging, not only for their wonderful appearance, but because they are so prickly on the leaves. They will keep an intruder out and animals in.

HONEYSUCKLE

(Lonicera)

FEATURES: Evergreen; height to 2 m; spread to 1 m. Non-invasive.

FLOWERS: Flared into two lips at the top of a tube. There are many kinds of lonicera, some known as honeysuckle, and the flowers, whilst generally similar, can be quite varied. *Lonicera japonica* is a rambler with sweetly perfumed, pink, yellow and white flowers. *Lonicera etrusca* 'Superba' has smaller flowers but the same shape, again perfumed, yellow with a flash of pink. *L. nitida* has tiny white flowers the size of a match-head, while *L. hildebrandiana* has yellow/orange flowers up to 15 cm long and perfumed. All loniceras are spring flowering.

LEAVES: Can also be quite variable. In *L. japonica* they are leathery and rounded, *L. etrusca* 'Superba' has a dark green, hairy leaf with a red underside, *L. nitida* has a tiny leaf to match its flower size, and *L. hildebrandiana* has large, soft, green leaves.

WHERE: There is not much of a problem with these shrubs and they will grow anywhere in Perth. *L. hildebrandiana* needs a hot spot as it is subtropical, whereas the others aren't. Give them a good, rich soil and regular applications of an all-purpose fertiliser as they are gross feeders. *L. nitida* needs a good watering regularly, more than the other varieties.

USES: Mostly they are grown over a fence or up a trellis and can be used as a screen or to hide something unsightly, like a mother-in-law. *L. nitida* is a bush and is used mostly in topiary or for small formal hedging.

COMMENTS: As a kid I can recall plucking the pretty flared flowers and sucking the sweet nectar out of the base. The birds and the bees know all about them too, and these lovely shrubs will attract them to your garden.

HYDRANGEA

(*Hydrangea*)

FEATURES: Deciduous; height to 2 m; spread to 2 m. Non-invasive.

FLOWERS: Vary greatly between cultivars. The large, round-headed balls of flowers most commonly grown are called hortensias — so named by a botanist for his mistress, Hortense! These flowers are sterile and range in colour from white through to a blood red, with every shade of pink you can think of. You could also say that they range from white through to the deepest blue, with every shade of blue you can think of — because these plants will do either, depending on whether the soil is alkaline or acidic! The lacecaps differ in that the flower heads are flat, with a ring of large sterile flowers surrounding an inner group of much smaller fertile flowers. Again, they range from pinks to blues depending on the soil type.

LEAVES: Usually quite large, up to hand size, and a deep green, almost milky colour. They can change to good autumn colour, especially in some species such as the Oak Leafed Hydrangea (*H. quercifolia*), which turns a very deep crimson.

WHERE: I am often surprised to see hydrangeas thriving by the seaside. For some reason it doesn't seem right, but they do! They also do well in the hills and all points in between. The 'hyd' part of hydrangea means water in Greek, thus they are water-loving shrubs. They prefer indirect light; an easterly or southerly aspect is best, and a well-drained rich soil.

USES: There are many varieties available now, from dwarf to medium and large shrubs. So you could plant some under a window, even if it's quite low, or on a fenceline to hide the old Super 6.

COMMENTS: The hortensias make a wonderful dried flower and can be dyed. The Climbing Hydrangea (*H. petiolaris*) makes its home in large trees such as oaks, and floats gorgeous lacecap flowers down a slender trunk. For a summer show it is hard to go past the hydrangeas and their beautiful huge flowers.

HANDY HINT

HYDRANGEAS. Hydrangeas don't require a massive prune back each autumn. It is worth dead-heading them and cutting back to a pair of plump buds, but other than that they are fine. There are many named varieties available now, some of which are dwarf — lovely under a low window. The colour range is huge but mostly among the blues and reds. A blueing tonic can be applied in late spring to encourage blue flowers. If you are 'into' hydrangeas, look out for some of the varieties that are quite different from the hortensias (these are the common varieties with the large round flower heads).

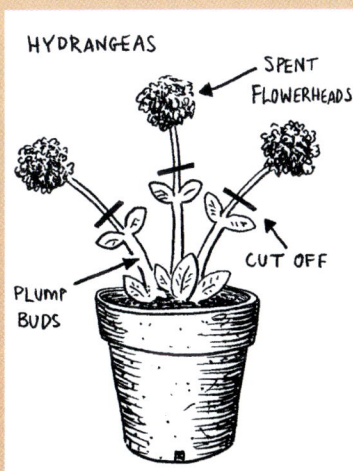

HYDRANGEAS

SPENT FLOWERHEADS

CUT OFF

PLUMP BUDS

ICE-CREAM BEAN TREE

(Inga edulis)

FEATURES: Evergreen; height to 3 m; spread to 2 m. Non-invasive.

FLOWERS: A little like an upturned shaving-brush of white anthers. These are quite attractive, but it is the seed pods for which they are grown. The fleshy pulp that surrounds the seeds tastes like vanilla ice-cream!

LEAVES: Deeply divided, dark green with brownish new growth. The leaf itself is quite large, about hand size.

WHERE: This is a tree in the tropics and a large shrub in Perth. It will not tolerate frost but does take cold, so be a bit careful in the hills. The coastal plain and coastal suburbs are fine. Find a part of the garden that is nice and warm summer and winter, give it some good tucker and water well over the hot months. An application of all-purpose fertiliser in spring and again at Christmas will keep it happy.

USES: This is grown in some countries as a food, but for us it will be for the yummy taste of the seed pulp. It is an attractive tree, too, with lovely summer flowers.

COMMENTS: This is a real winner with the kids, and a good way to introduce them to the excitement of gardening. It may take a couple of years to fruit — but worth the wait.

INDIAN HAWTHORN

(*Rhaphiolepis*)

FEATURES: Evergreen; height to 1 m; spread to 50 cm. Non-invasive.

FLOWERS: *Rhaphiolepis x delacourii* has heads of fragrant pink flowers about the size of a ten cent coin. *R. indica* produces white and pink flower heads, while a cultivar, 'Springtime Spring Song', has pink flowers with a white centre. *R. umbellata* has white perfumed flowers with prominent red stamens. All are spring flowering.

LEAVES: Tough and leathery, making them quite resistant to salt spray and the coastal breezes. The edges are serrated and the colour is a sea-green.

WHERE: They will tolerate a poor soil, but if it is built up a little the flowering will be better. They are also quite drought-tolerant, once established. They will grow from the seaside to the hills.

USES: A good plant to mix in with silvery foliaged plants in a coastal garden, as they will add a green foil to the balance. The flowers are surprisingly pretty and delicate for such robust shrubs. They will grow for quite a while in a container before you have to replace the potting mix. On their own in any garden, they are a handsome shrub.

COMMENTS: Almost indestructible, they are excellent for a beginner gardener. The fragrance is very lovely. Sometimes called the Indian Hawthorn, but in fact they come from Japan and Korea, which goes to show that common names are sometimes inaccurate.

INDIGO BUSH

(Indigofera decora)

FEATURES: Deciduous; height to 75 cm; spread to 50 cm. Non-invasive.

FLOWERS: Wisteria-like, pink and white, hanging from this low shrub in summer. These racemes are very pretty, with the papery flowers lasting through to autumn. *I. australis* is an evergreen form with typical pea flowers, also pink, hanging in racemes.

LEAVES: Oval, coming to a point, a soft green. The overall appearance is quite delicate.

WHERE: This is a tough little shrub, very easy to grow. Anywhere from the coast to the hills is fine, in a rich, well-drained soil. Water well to establish, but they have proven drought resistance — although as in most things a good water over summer will keep them looking much better and give a better show of flowers.

USES: Ideally suited to cottage gardens, the plant will divide by short underground rhizomes. After a few years the entire shrub can be dug, in mid-winter, and these cut off for new plants. They are quite at home in a large container for a patio. The dye indigo is extracted from the root system (if you are into natural dyes).

COMMENTS: This is another plant that should be grown more widely. Great for the novice gardener because it is hard to go wrong with it. I find the extended flowering season a boon when other things are going out of flower.

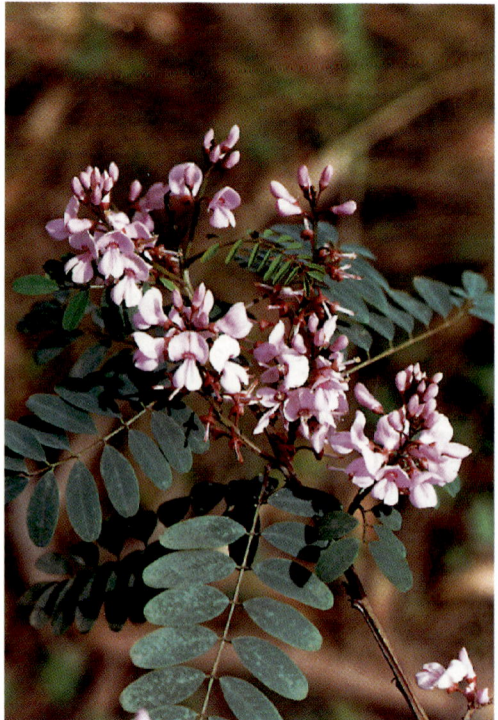

JAPANESE PLUM-YEW

(*Cephalotaxus harringtonia*)

FEATURES: Evergreen; height to 2–3 m; spread to 1 m. Non-invasive.

FLOWERS: Inconspicuous.

LEAVES: Like a yew, long, thin and a dark matt green. They are almost strap-like, about 30 mm long and 5 mm wide. They are carried fairly heavily on the bush and create a pleasing look.

WHERE: This shrub can be grown anywhere other than the seaside. The cold of the hills won't bother it, neither does the heat of the coastal plain. A good, compost-rich soil and an annual feeding with good watering over the hot months are sufficient.

USES: The Japanese Plum-Yew is upright in its habit and is used for best effect in an architectural setting. It fits well with formally arranged gardens, and although it can be clipped, it rarely needs it. A pair at the entrance of a pathway looks stunning, and an avenue lined either side with them is a knockout.

COMMENTS: Although it is classified as a conifer, as are yews, it doesn't have the nasty habit of dying suddenly, as the Irish Yew does. So if you are growing a row of them it is very unlikely that, after ten years, you will have a gaping hole. The naturally upright growth makes a very bold statement in a well-designed garden.

LAVENDER

(Lavandula)

FEATURES: Evergreen; height to 1 m; spread to 75 cm. Non-invasive.

FLOWERS: Have an unmistakable fragrance and colour, renowned around the world for centuries. The flower colour ranges from white to mauve. There is a bluer form, *L. canariensis*, and a quite blue form in its Australian-bred cultivar 'Sidonie'. The flowers are oily — if you crush them the pungent fragrance will remain on your hands for some time.

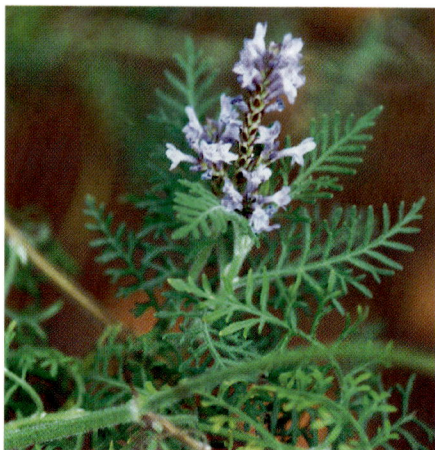

LEAVES: In most species are generally similar, sage-coloured and indented on the margins. They vary in size and shape according to the species, but the most significantly different is *L. canariensis* which has a fine, fernlike leaf and is green.

WHERE: Lavenders are at home by the seaside, growing in poor soil with little water. Sound like your place? All you have to do is reproduce those conditions and they will be at home. They will grow anywhere in Perth, and are just as at home in the hills as by the coast. Just lay off the water, other than to establish, and give them full sun.

USES: There is no sight quite like a lavender field where the flowers are harvested for the essential oil. In full flower and covering many acres, it is spectacular. The lavenders can be used for a low hedge (very good in Tuscan gardens), or free-standing in a cottage garden. They are very nice mixed with other silvery-leaved plants in a seaside garden.

COMMENTS: We clip our hedge once a year and for the rest of the time it looks marvellous, such an easy plant to grow. If you find yours are just not performing it may be a lack of dolomite lime. This contains manganese, which they do need, so throw a couple of handfuls around them and see what happens — I'll bet it works! The only other problem you may run into is humidity; they dislike it intensely, except 'Sidonie' which was born in Sydney so it must like it!

LILAC

(Syringa)

FEATURES: Deciduous; height to 3 m; spread to 1 m. Non-invasive.

FLOWERS: Rather hyacinth-like: large bunches of florets gathered together on a spike that rises above the foliage. The colour range is large, about thirty variations from pure white to the deepest purple. You can find soft pink, soft blue, violet, mauve and red. These glorious blossoms are also sweetly scented.

LEAVES: Dark green and oval, about the size of a small egg. Sometimes they produce a bronzy-red autumn colour.

WHERE: Best in the hills. Lilacs prefer a cold autumn and winter to set flowers the following spring. I have seen some good specimens in flower on the coastal plain, but these are the exception rather than the rule. A rich composty soil with a neutral ph, a deep mulch to keep the roots cool and moist, and good summer watering are the basic necessities.

USES: As a beautiful spring show accompanied by that sweet fragrance. They will grow quite bushy and produce suckers: both these and the old wood will carry the new season's flowers. For this reason they need some room and will provide a good screen or hedge.

COMMENTS: Syringa is the only grafted shrub or tree that needs to be planted with the graft below the soil. The shrub is grafted onto Ligustrum as a 'nurse' graft which is designed to keep it alive until the lilac produces its own roots. If you don't, when the short-lived Ligustrum dies, so will the lilac. If you have one planted with the graft above the ground, dig it up in winter and plant it deeper.

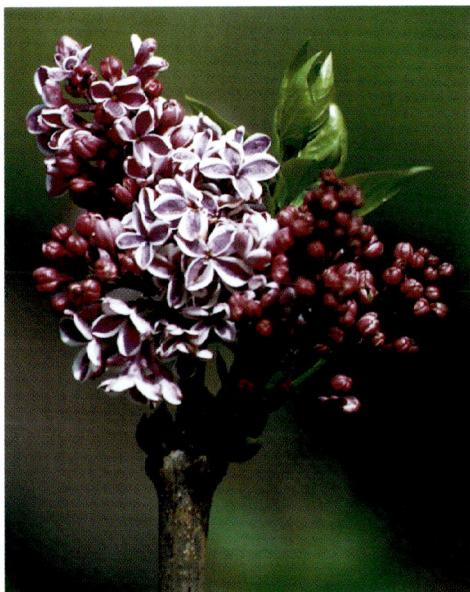

LILY-OF-THE-VALLEY

(Clethra arborea)

FEATURES: Evergreen; height to 2 m; spread to 1 m. Non-invasive.

FLOWERS: Dainty little bells that hang downwards in a raceme, each bell about the size of a pea. They are highly perfumed and can appear anytime from summer until autumn, although I have seen them flower in spring. The flowers are mostly white with a pretty pink blush.

LEAVES: Large, an elongated oval with a serrated edge, hanging downwards.

WHERE: Clethras love a cool acidic soil and take a dappled sunlight position quite happily; they don't like hot winds. Soil can be acidified with camellia fertiliser or an acidifier, and should be compost-rich. Best grown in the hills or the coastal plain, but not by the coast.

USES: Use a clethra as an elegant addition to the garden. They are rarely seen in Perth but do well if the conditions are right. They would look wonderful in a bed of *Pieris japonica*, which has similar flowers and likes.

COMMENTS: The flowers are very special, running out along a stem and just hanging there, offering their delicious fragrance.

HANDY HINT

WORMS. Worm farms are a brilliant way to recycle your veggie scraps. Usually designed on a rotating 2 - 3 tier system, the worms eat their way up from the bottom to the top. The worm wee is a great gentle fertiliser either straight or diluted and the casings are an excellent addition to potting mix. The handiest, mess free pets imaginable, and they'll even eat your old newspapers. Some of your more enlightened local councils will sell rate payers a farm at subsidised prices. If your council doesn't, lobby them until they do.

MAHONIA

(*Mahonia*)

FEATURES: Evergreen; height to 2 m; spread to 1 m. Non-invasive.

FLOWERS: Produced in racemes of many flowers, always yellow. In the case of M. *lomariifolia* these sprays are on the end of a branch. *Mahonia aquifolium* will produce more of a panicle of golden bells looking like a bunch of golden grapes, hence its local common name of Oregon Grape. The flowers of both species are very showy and fragrant, appearing from late winter to spring.

LEAVES: Spiky, dark green and resembling holly. They are leathery and able to withstand frost and wind. A distinct advantage is that the prickles make it difficult for people to get in and cut the flowers, so you are always assured of a good flower display.

WHERE: Good in the hills of Perth and on the coastal plain, but not by the coast. A rich composty soil with some well-rotted poo and good watering over the hot months are required. These shrubs will enjoy a deep mulch and a bit of room to move, preferably where you don't have to brush past them.

USES: Just as ornamentals. When they are out of flower, the attractive foliage, arranged in whorls, is an interesting highlight.

COMMENTS: It is quite a spectacle when mahonias are in flower; the heads of blossom are a dominant feature in the garden. They can look wonderful if grown under the canopy of a large open tree.

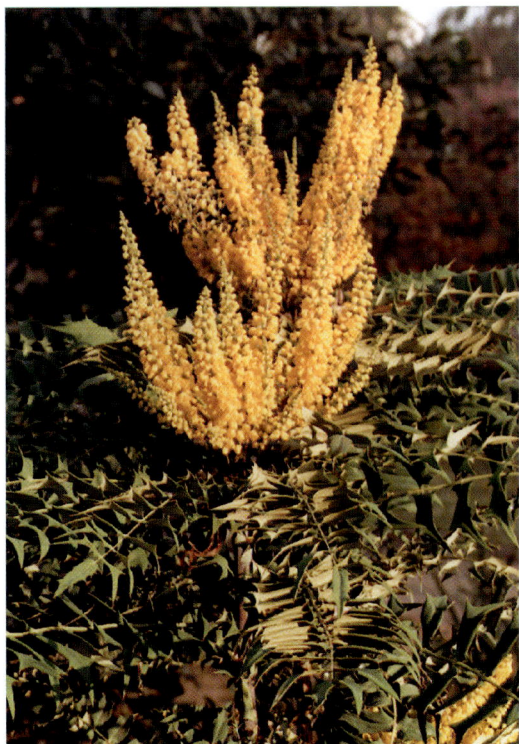

MALLOW

(Lavatera acerifolia)

FEATURES: Evergreen; height to 1.5 m; spread to 1 m. Non-invasive.

FLOWERS: A very rich deep pink open flower, like a small hibiscus (to which they are related). An inner band of deeper pink runs around the petals just adding to the loveliness. The flowers begin in late winter and last through to at least mid-summer.

LEAVES: Dark sea-green and shaped like a maple, hence its name 'acerifolia' or 'maple-like leaf'. The leaves on an individual shrub vary from almost apple-sized to cumquat-sized and make a distinct display all on their own. When the flowers are added they are offset superbly against this foliage.

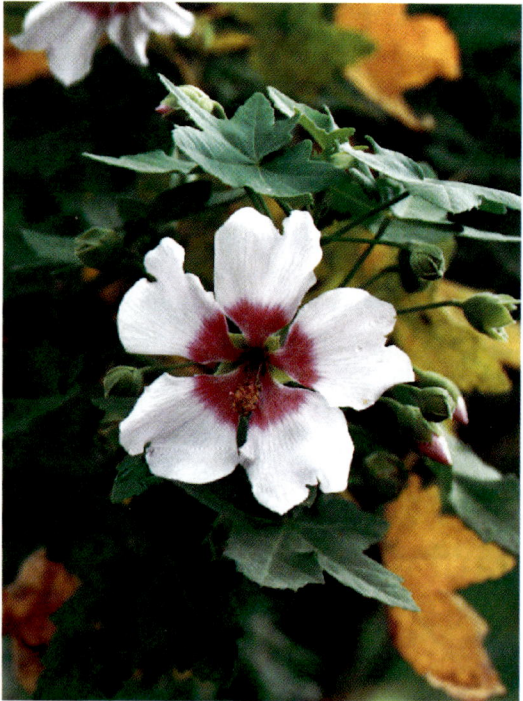

WHERE: Highly adaptable to all conditions, they will grow anywhere in Perth away from violent winds. Soil requirements are not a lot, a bit of enrichment and light watering over summer.

USES: Almost a requirement for a cottage garden, they will add a lovely splash of rich colour to any garden. A massed planting is a delight and, if used for hedging and lightly pruned, the flower show all over the hedge will be glorious.

COMMENTS: I have a great joy in seeing such magnificent blooms on such neglected plants as ours are. We should be seeing a lot more of this very-easy-to-grow shrub. A bit of dead-heading will prolong the flowering period.

MATILIJA POPPY

(Romneya coulteri)

FEATURES: Evergreen; height to 2 m; spread to 1 m. Non-invasive.

FLOWERS: Large, poppy-shaped, with pure white crepe petals and a large centre of yellow stamens. A striking flower, it can be up to 20 cm across and deliciously perfumed. Flowering is over spring and they can be used for cut flowers, being long-lasting.

LEAVES: Deeply dissected and a sea green colour. Most attractive foliage, although not a lot of it.

WHERE: Grow them in a poorish soil, water occasionally and deeply, and a bit of mulch around the base. Plant them in dappled light but near sunlight, and let them grow towards the sun. When you plant them, make sure it is in an area that won't be disturbed — this shrub deeply resents root disturbance so much that it will promptly die! Romneya is a native of California and will thrive in Perth.

USES: As an ornamental in a quiet part of the garden. The flowers are really quite something, and when you are wandering around in the little-visited part of the garden you will be thrilled to find it in full flower.

COMMENTS: Devilishly difficult to propagate, almost impossible from seed, it is necessary to take root divisions in autumn after cutting all the growth back to the ground. This process usually results in the death of the parent plant! If you can find one it won't be cheap — but then neither are diamonds!

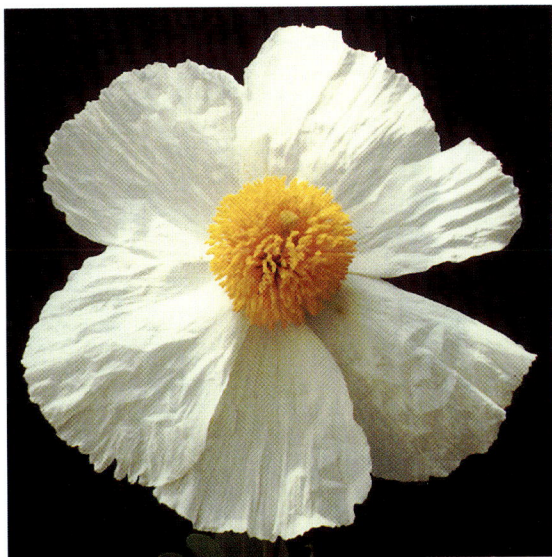

MICHELIA

(*Michelia*)

FEATURES: Evergreen; height variable to 4 m; spread to 2 m. Non-invasive.

FLOWERS: A range of colours from very showy snow white to creamy non-descript and deep pink. They all have one thing in common though — absolutely delicious fragrance. The Portwine Magnolia (*Michelia figo*) has small, egg-shaped flowers often hidden in the foliage, beginning cream and opening to a wine colour, with a perfume to match, just like the sweet smell of port. *Michelia doltsopa* produces large white flowers that open like a *Magnolia stellata* and exude a gorgeous scent, almost lemony and sweet. *Michelia champaca* has a small cream, rounded flower, not all that prominent, but boy, the fragrance! This is my favourite, very hard to describe but a little like a banana cream mixed with true vanilla.

LEAVES: In M. *figo* the leaf is dark green, oval and small, whilst M. *doltsopa* has very large, milky-green leaves that cover the whole bush. M. *champaca* has a small rounded leaf, with a khaki colour in new growth.

WHERE: M. *figo* can be grown anywhere away from frost, preferring a warm position in summer and winter, as does M. *champaca*. M. *doltsopa* likes a cooler climate and will do well in the hills and cool parts of the coastal plain. They all enjoy a well-drained soil enriched with rotted animal manure, a deep mulch and very good watering over the hot months.

USES: Don't grow them too far from the house as you will want to enjoy the perfume as often as you can. We have two M. *figo* near our backdoor, and the perfume is there for months at a time. M. *doltsopa* can be used as a screen, but I would have all of them in a prominent position as ornamentals.

COMMENTS: When it comes to describing the perfume that nature offers us, words are very inadequate, especially when the perfumes produced by this group of magnolia cousins are just so sumptuous. I think you really need to head off to a good nursery when they are in flower and see what I mean.

HANDY HINT

SWEET PEAS. Sweet peas are a delight in spring and the best way to maximise this easy-to-grow annual is to give them a dose of lime when planting. It is a very good idea to set up stakes or a trellis before you plant the seed. Once they make a move out of the ground they really take off, and trying to shove in a stake later can leave them looking unsightly or with damaged roots.

STAKE SWEET PEAS

SET UP TRELLIS OR STAKES BEFORE PLANTING.

MINT BUSH

(Prostanthera)

FEATURES: Evergreen; height to 1 m; spread to 1 m. Non-invasive.

FLOWERS: *Prostanthera rotundifolia* has mauve bells held on an upright stem, massed along the branches and terminal. *P. ovalifolia* is completely covered in pinky/mauve bell-like flowers, and *P. ovalifolia* 'Alba' is the same with white flowers. *P. cuneata* 'Alba' has a flower a little like a Geraldton Wax but it is soft and white. All prostantheras flower over spring.

LEAVES: Vary from rounded ('rotundifolia'), to an oval shape ('ovalifolia'). The colour is a light green and they are all aromatic, giving a mint fragrance when crushed.

WHERE: Not at all specific in their requirements and, being an Australian native, a well-drained, not overly rich soil and sporadic watering are all they need. They will grow anywhere in Perth.

USES: Looking as though they belong in a cottage garden, the early settlers often used them for this purpose. A row of prostantheras can make a wonderful spring show, just as one or two dotted around the garden will do the same. They will grow well in a container with the odd clip to maintain the shape.

COMMENTS: These are so easy to grow that they would make a top gift for a black-thumb, or a novice gardener. The spring show is brilliant but brief, but the foliage is so attractive that they will always look good. They are lovely in a herb garden, too!

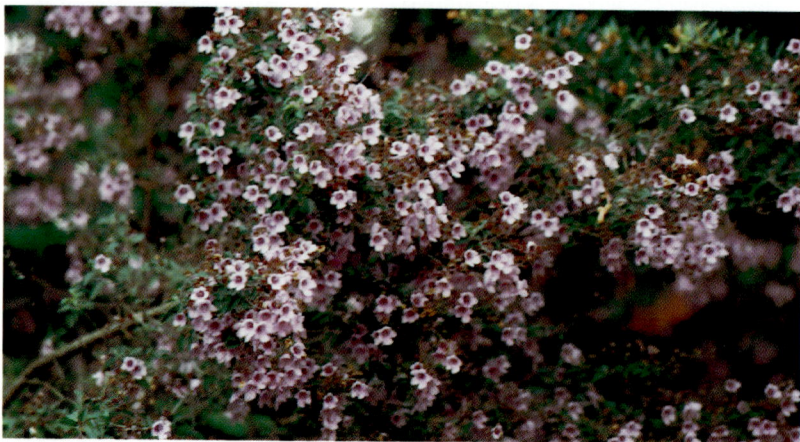

MOCK ORANGE

(Philadelphus)

FEATURES: Deciduous; height to 2 m; spread to 1 m. Non-invasive.

FLOWERS: White, bell-shaped, with a yellow centre of stamens, hanging along the length of long canes. The weight on the canes pulls them down to nose height, which is ideal to take advantage of the heavenly fragrance. Flowering from spring onwards, often after other spring-flowering shrubs are finishing. *Philadelphus x lemoinei* is a hybrid with lovely ruffled blossom, whilst *P. x virginalis* is another hybrid, of uncertain parentage (could this be the reason for its name?) that produces double white perfumed flowers. *P. coronarius* has fragrant bells that hang on a semi-evergreen shrub.

LEAVES: Hairy or felted, soft green, elongated, coming to a point. The colour of the leaves is quite attractive when out of flower and, being carried on canes, adds a new dimension to the garden.

WHERE: Philadelphus belong to the hydrangea family and, like hydrangeas, are quite happy in either alkaline or acidic soil. Like hydrangeas, they like a good regular watering, a rich composty soil and some all-purpose fertiliser in spring and again at Christmas. Grow them away from violent winds, from the coast to the hills. It is a good idea to cut them back hard over winter to encourage new spring canes that will carry next season's flowers.

USES: Only as an ornamental, they can be mass planted for a stunning spring show, or plant one near a pathway so the canes cascade over and when you walk past you can enjoy the fragrance.

COMMENTS: What I love about these delightful shrubs, apart from the rich perfume which can smell like orange blossom, is the white of the flowers. They seem to be very pure, like milk. They are really easy to grow and will suit any garden.

MYRTLE

(*Myrtus*)

FEATURES: Evergreen; height to 3 m; spread to 1 m. Non-invasive.

FLOWERS: White and similar to cotoneaster flowers. They vary little between species. M. *communis* has five petals, whilst M. *luma* has four. After flowering, small black berries are produced, much prized by birds, usually in autumn after the summer blossom.

LEAVES: Aromatic, rounded, coming to a sharp point, glossy and dark green. They are held in profusion over the bush and usually hide the branches and stems.

WHERE: They will grow from the hills to the coastal plain, M. *communis* going all the way to the coast. Slow growing they like a rich, well-drained soil and good watering to establish, after which they show some drought tolerance.

USES: Primarily as hedging shrubs. Their slow growth keeps them in check and looking neat in a formal hedge. M. *luma* is now being used extensively for topiary, being the faster growing of the two. The next time you see a topiary giraffe or elephant, chances are it is M. *luma*. This very useful shrub is perhaps the most widely grown in topiary circles, if you will pardon the pun.

HANDY HINT

LAVENDER. Lavenders are most resilient plants that tolerate drought, seaside air, limestone and almost anything you or nature can throw at them. Sometimes, however, they go downhill for no apparent reason and look horrible. Nothing you do seems to help. Well there is probably one thing you haven't done — top dress with dolomite lime! Lavenders need lime, but more importantly they need occasional doses of magnesium. Dolomite is a naturally-occurring compound of both, and a good handful thrown over a bed of lavender will do wonders.

NUTMEG BUSH

(Tetradenia riparia [Iboza riparia])

FEATURES: Deciduous; height to 1.5 m; spread to 1 m. Non-invasive.

FLOWERS: In mid-winter with a long panicle of massed white, tiny flowers, tinged with blue from the purple anthers. The overall effect is just lovely and the fragrance is very spicy.

LEAVES: Very soft, velvety, light green. The leaves are serrated on the edge and quite rounded, coming off a fleshy stem, a bit like a geranium.

WHERE: Tetradenias are a tough shrub without many needs. Average soil with an all-purpose fertiliser in spring, and water once or twice a week over summer. They can be grown easily from the coast to the hills.

USES: They will grow tall enough to be a screen, but are best if reduced heavily after flowering to keep them compact over summer. Lovely in a cottage garden, they will add an interesting contrast to other shrubs with their winter flowering and different foliage. They will fit into a herb garden nicely, giving off their spicy perfume.

COMMENTS: A few years ago I planted some in a windbreak area and left them to their own devices, and now every winter we have this wonderful display for no effort. They are excellent for the beginning gardener who will be encouraged by the ease of growing them. Grows readily from cuttings.

OCHNA

(Ochna serrulata)

FEATURES: Evergreen; height to 2–3 m; spread to 1 m. Non-invasive.

FLOWERS: Pretty little yellow/orange bells that hang downwards off the branches. These give way to a calyx that swells and finally opens to reveal the seeds. The calyx itself is tomato red and the four seeds are arranged in perfect symmetry on the calyx — they are jet black — and it looks for all the world like a miniature Mickey Mouse hanging on the plant. The effect is quite startling and I have seen nothing else quite like it.

LEAVES: Bronze new growth followed by a gentle green, the leaves are very serrated and wavy. They are long, like a strap coming to a point.

WHERE: This shrub likes dappled sun, so under a large open tree or shade cloth is ideal. Ochna likes a rich soil and good watering over summer. It can be grown happily in a large container.

USES: A good plant to win bets with your friends. Tell them you have a shrub with Mickey Mouse hanging on it and they won't believe you — heh! heh! It really is a most attractive, almost two-dimensional shrub that will look handsome in any garden.

COMMENTS: This again is a good one to help interest the kids in gardening. I love it myself, but maybe I'm just a big kid!

ORANGE JESSAMINE

(Murraya paniculata)

FEATURES: Evergreen; height 2 m; spread to 1 m. Non-invasive.

FLOWERS: In a creamy panicle, decorating the shrub all over. A generously flowering shrub that has a perfume like jasmine and flowers like orange blossom. Enticing you into the garden to smell the fragrance in early summer, it will repeat the performance a couple more times in one season.

LEAVES: Pinnate, dark green and dense, making a superb backdrop for the flowers. The leaves are about the size of an egg with an attractive gloss.

WHERE: Murrayas enjoy a humid climate, although they are quite at home in Perth. They do not like frost, but will take cold well. Try them in the hills and down to the coast, avoiding wild winds. By giving them a deep mulch you can help increase the humidity, and feeding them with an all-purpose fertiliser regularly will keep them lush and green. Water well over summer.

USES: Now being used more and more for hedging, it is becoming more widely known. The flowers, carried on a neatly clipped hedge, are a bonus. Very good for screening wide areas or as a stand-alone with its mass of heavy blossom over a long period.

COMMENTS: M. *alata* is another species that doesn't perform as well as M. *paniculata*, but is easier to grow from seed. It is commonly sold in nurseries, but in my opinion it is worth hanging out for M. *paniculata*.

OSMANTHUS

(Osmanthus fragrans)

FEATURES: Evergreen; height to 2 m; spread to 1 m. Non-invasive.

FLOWERS: Tiny, white, insignificant except for one thing – the fragrance. This must be one of the most delicious perfumes in the world. It is addictive; once you have experienced the perfume of O. *fragrans* floating on the wind, you will know what I mean – you will just have to have one. Osmanthus will let you know it is in flower by its perfume which is a little like crushed apricots mixed with jasmine. Flowering is from autumn until spring.

LEAVES: Dark green, a bit like camellia leaves, serrated and leathery. They are profuse, covering the bush, and helping to hide the flowers.

WHERE: They must have a free-draining, rich soil built up with well-rotted manure. Give them a good watering over hot months until established, which may be a couple of summers. They can be grown from the hills to the coast.

USES: Use them as a rough and irregular screen, but keep them within reach of your nose! They are not all that attractive other than the fragrance.

COMMENTS: It is hard to imagine that such glorious perfume can come from a member of the olive family. There are a number of less fragrant varieties, but why bother! Except for O. *heterophyllus* 'Variegatus', which is a variegated form with a quite respectable fragrance.

OSMOXYLON

(Osmoxylon lineare)

FEATURES: Evergreen; height to 1 m; spread to 50 cm. Non-invasive.

FLOWERS: A small clump of white flowers held high on a single spike.

LEAVES: Frond-like, being dissected and arranged in whorls on the trunk. They will reach a spread of around 50 cm, dark green and with a matt surface. The whole effect of this shrub is that of a miniature palm. An ideal indoor plant.

WHERE: Growing naturally in a jungle atmosphere, with overhead shade from large trees, this small shrub will fit easily into a tropical-look garden. Keep it away from direct sun, give it a rich soil and water it well. Under shade cloth or on a patio will be fine. Not liking the cold, it is best grown on the coastal plain, including the coast itself, in a protected position.

USES: Osmoxylon is like a little version of a palm and will fit in well as a low shrub that will help fill an area that is calling out for something tropical but small. It is just as happy grown in a pot or container and used as an indoor shrub. It needs light but not the direct sun coming through a window.

COMMENTS: An attractive plant to fit into an awkward area, it is best fed with a slow-release fertiliser (8–9 months) and well watered. As an indoor plant it can be grown well in the hills, moving it out to a sheltered position once the cold has gone.

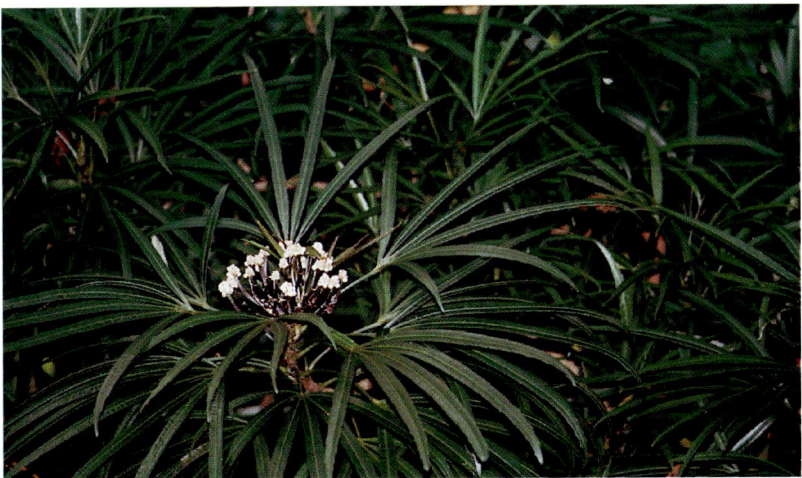

PHOTINIA

(Photinia glabra)

FEATURES: Evergreen; height to 4 m; spread to 2 m. Non-invasive.

FLOWERS: In broad panicles of flat 'heads' composed of many hundreds of white flowers. The flower heads can be as large as a hand and appear mainly in spring, although it is not unusual to have a flush of flowers in autumn and early winter.

LEAVES: Large, very dark green. In autumn, some of these leaves turn a brilliant scarlet as the shrub will shed a proportion at this time. The new growth is spectacular, in the case of *P. x fraseri* 'Rubens' a luminous red, while *P. x fraseri* 'Robusta' has bronzy red new growth.

WHERE: Photinia will grow well in Perth from the hills to the coast. Traditionally, they are a cold-climate shrub that has adapted well. They love a rich acidic soil, good mulching to keep the roots cool and moist and will grow happily in partial shade to full sun. Keep them well watered and enjoy the show.

USES: One of the most popular of all hedging shrubs. Easily kept under control and marvellous when in full flower, literally covering the hedge with masses of white blossom. The clipping required to maintain the hedge allows all that rich red new growth to cover the hedge at other times. The density of the growth will make the plant impenetrable, both to see and walk through. Also used to hide unsightly areas, like a submarine base.

COMMENTS: A new hybrid has recently been released onto the market. It was bred in Sydney and is called 'Superhedge'. This is a far more vigorous shrub, with the same wonderful characteristics as other photinias, except that it grows twice as quickly. I suspect that this will be the best hedging plant of the future. *P. beauvardiana* is a deciduous species that changes to the most spectacular range of reds and crimsons before shedding its leaves.

HANDY HINT

BEWARE LABEL TIES. When you buy a tree or shrub that has a tie-on label rather than a pictorial label, you will notice that it is tied with what appears to be cotton thread instead of the green twist-tie that accompanies pictorials. This is, in fact, nylon so it will last. However, if it is tied around a branch or, worse, the trunk, it will not expand, but instead can ringbark the branch or trunk. Best to take it off and put it on a small stake next to the plant, or remove it altogether.

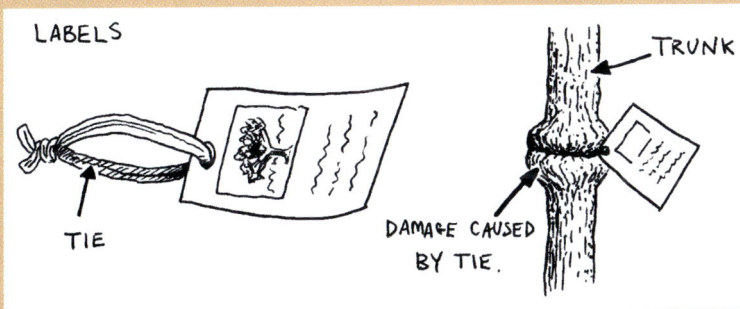

LABELS

TIE

TRUNK

DAMAGE CAUSED BY TIE.

PHYLLANTHUS

(Phyllanthus minutiflorus)

FEATURES: Evergreen; height to 1 m; spread to 50 cm. Non-invasive.

FLOWERS: In a spray of tiny pink blossom in spring, followed by little pea-like seed pods.

LEAVES: Only 1 cm long, carried on clustered branches and are very, very densely arranged on this shrub. A dark green colour, the effect of the growth is like a waterfall of tiny leaves. The individual leaf shape is elongated, and the texture is soft and ferny.

WHERE: Avoid frost-prone areas, but otherwise they will grow well all over Perth. A rich, well-drained soil with compost and cow manure added, and a regular feeding with an all-purpose fertiliser accompanied by good watering, will produce a wonderful bush.

USES: Phyllanthus have the most marvellous waterfalling effect. The many slender branches, just covered in leaves, rise from the centre of the shrub and fall over one-another. They will add a very dramatic effect to a terraced wall when they are grown by the edge and allowed to cascade over it to the ground below. In a hanging basket, the long trailing branches will add an accent to an area. Used as a ground cover, Phyllanthus will spread to a good metre and create a mounding effect.

COMMENTS: I have seen this useful shrub used as a low hedge and it looked stunning. Although not commonly grown, there is no reason why it shouldn't be used somewhere in all gardens.

POINSETTIA

(Euphorbia pulcherrima)

FEATURES: Deciduous; height to 2 m; spread to 1 m. Non-invasive.

FLOWERS: Interestingly, are quite insignificant for such a spectacular plant. The vibrant display we see as winter draws nigh is a collection of modified bracts around the small yellow and green flowers. These are produced as the days shorten and the colour is heightened by the cold weather. *E. pulcherrima* is the poinsettia, and is now available from dwarf size to medium shrub and quite tall shrubs. With hybridisation we also have differing colours in these bracts — pink, white and red and the fabulous deep crimson for which they are famous. *E. pulcherrima* 'Henrietta Ecke' has curiously curled leaves and produces a brilliant ball of colour instead of the flat flower 'head' of the other types.

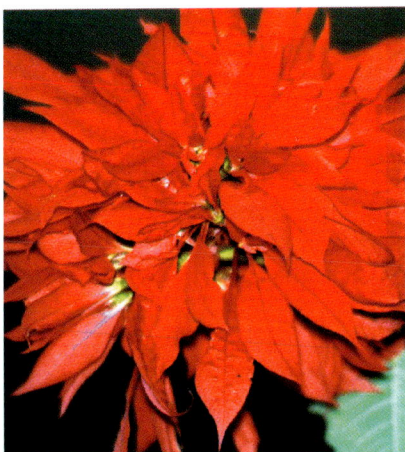

LEAVES: As big as your hand, elongated and ribbed with red veins, commonly with new growth having tinges of pink or red. The leaves have a broad serration and a wave in them.

WHERE: Anywhere that is frost-free. These shrubs are native to Mexico and like a rich soil with plenty of summer watering. Once established, they show some drought tolerance but will always look better if they are watered well. By the coast it is advisable to grow the dwarf or medium-height variety to avoid wind damage. Put them in a hot spot in your garden and wait for winter.

USES: Primarily as ornamentals. Planted in a group, particularly with different cultivars, they can look sensational.

COMMENTS: The indoor poinsettias are forced by clever manipulation to flower out of season, generally at Christmas time. These plants are really grown to have a life a little like a bunch of flowers — once the show is over, throw them into the compost bin. You can keep them going, but your really need the magic of the grower. With the garden-grown varieties it is best to reduce them by half after flowering.

RHODODENDRON

(*Rhododendron*)

FEATURES: Height to 3 m; spread to 2 m. Non-invasive.

FLOWERS: Very large bunches of flowers tightly arranged on terminal growth. Often trumpet-shaped and ranging widely in colour, shape and size. Azaleas are in fact small rhododendrons, again with a myriad collection of sizes and colours. Considered to be the most beautiful of all the flowering shrubs with a vibrancy of colour and profusion that is breathtaking. There appears to be no limit to the colour range — white, pink, yellow, orange, blue, cerise, red, scarlet, crimson — and then every variation in between. There are over 800 species and several thousand hybrids! Flowering is in spring.

LEAVES: Thick, large, ovate, dark green. They tend to clothe the plant over the perimeter making them attractive even out of flower.

WHERE: The conditions required by this group are quite specific. No hot drying winds, but dappled sunlight, quite acidic moist but well-draining soil, and a cool to cold winter. There are some splendid specimens in Araluen Botanic Park, many years old and massed with flowers in spring. A regular mulching is needed and well-rotted cow manure is a definite advantage to growing them on the coastal plain. They do best in the hills, and if you live by the coast the best you can hope for is a photograph of a rhododendron in your garden! They can be grown as a potted specimen for a while.

USES: Mainly as a dazzling display of colour, nice massed with smaller azaleas around the undercanopy.

COMMENTS: Although there are many thousands of rhododendrons, only about thirty or so are hardy enough for the Perth climate. It is worth hunting them out and talking to an experienced nurseryman at a specialist nursery. The Rock or Vireya Rhododendrons are easier to grow, coming from the highlands of New Guinea. The flowers are smaller, more tubular and the plant more on the small side. These are being successfully hybridised and some very vibrant colours are becoming available.

HANDY HINT

ROLL CALL. This next little hint may save a few trees from a fate worse than death. Before popping your toilet roll onto the holder, tread on it first so that the little cardboard centre forms an oval. When you pull the paper off the roll it won't spin out a metre of paper, but about a quarter of a metre. More than enough for one application! Your kids may moan but they need something to whinge about.

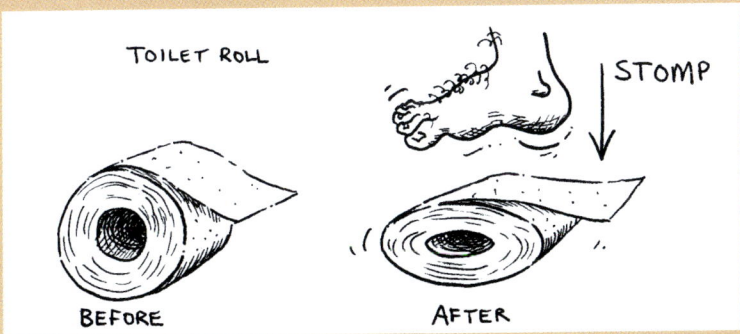

TOILET ROLL

STOMP

BEFORE

AFTER

ROCK ROSE

(Cistus)

FEATURES: Evergreen; height to 0.5 m; spread to 0.5 m. Non-invasive.

FLOWERS: Five-petalled, open and daisy-like. The colours are white, off-white, pink, white-pink and cerise, all with yellow stamens in the centre of the flower. Some varieties have a crepe-like ruffle to the petals.

LEAVES: Either a soft green or sage-like, tough and aromatic.

WHERE: For all those poor people who live by the beach, this is the plant for you. Cistus likes poor soil, little water and no humidity. The salt spray doesn't bother them, nor does the limestone. Having said that, they also do well in the hills and all points between.

USES: Ideal to spot around the garden for the lovely flowers and long flowering period, and of course their ease of care! A little clipping to keep them compact is about all they need and then you get the lovely aroma from the leaves.

COMMENTS: Good in a rockery or with the differently coloured varieties massed together, they are, it seems, designed to grow where nothing else will.

HANDY HINT

SECATEURS. Secateurs are the most important tool a good gardener can have. When you buy a pair, be prepared to spend some money. Chefs will use only carbon steel blades, and likewise good secateurs have a carbon steel blade. It simply means that they can be sharpened on an oilstone. Blunt secateurs do more damage than good and actually crush the plant rather than cut. Don't be tempted with an 'el cheapo' pair, you are just throwing money away. Always sharpen them before use and sterilise the blade in alcohol or methylated spirits before and after use.

SNOWBUSH

(*Breynia nivosa*)

FEATURES: Evergreen; height to 1 m; spread to 50–75 cm. Non-invasive.

FLOWERS: I bet you can't find them! They are very small and insignificant, although I think they are quite charming. Happy hunting.

LEAVES: Very colourful. Breynia has a leaf that is oval in shape and in three colours. These colours are not a distinct variegation but mottled. The same leaf may have white, pink and green in various shades. The older and lower leaves tend to be dark green.

WHERE: Breynia has been grown all over the place in Perth, from high up in the hills all the way to the coast. It is found mainly in the older suburbs such as Subiaco, Victoria Park and Guildford. Best grown in a warm sunny spot. Water well in summer for the best effect, although it has proven quite drought-tolerant. Give it some rotted manure and a deep mulch and you will be very proud of your gorgeous shrub.

USES: Mainly as an ornamental, although I have seen a couple of stunning hedges. An old favourite in cottage gardens for foliage contrast — they make a good background shrub.

COMMENTS: The full cultivar name, *Breynia nivosa* 'Rosea-Picta', is a little off-putting and most people won't know it, though they will recognise the shrub. It is one of the loveliest of colour combinations. It will take part-shade but the colours won't be so strong. The plant divides by suckers, but as they mainly come up by the parent bush they rarely present a problem. A rare cultivar called *B. nivosa* 'Iron Range' has coppery-brown and khaki-green leaves; sounds horrible but it is very attractive.

SPINDLE-TREE

(Euonymus fortunei)

FEATURES:
Evergreen; height to 2 m; spread to 1 m. Non-invasive.

FLOWERS:
Insignificant, tiny and white.

LEAVES: Grown as a foliage specimen. There are a number of different cultivars. 'Emerald Gaiety' has variegated leaves of green and white about the size of a small egg, with a similar oval shape. 'Silver Pillar' has a pronounced white margin on a dark green leaf. 'Emerald 'n' Gold' has an attractive gold variegation on the leaf margins. All varieties have a serrated leaf edge.

WHERE: Try to avoid wild winds, but otherwise these hardy shrubs will do well in Perth, from the hills to the coast. They prefer a compost-rich soil, and good watering until established.

USES: Often used as a low hedge clipped to a very formal shape, they look stunning with the leaf variegation shown off well. Clipping encourages tight new growth that will cover the bush. Because of the foliage, euonymus can be used to great effect in a rockery or as a single specimen, or even used with a few varieties placed adjacent to each other.

COMMENTS: Easy to grow and maintain, yet such an effective contrast can be had by using the richly coloured foliage. Originally from Japan, new cultivars appear from time to time and it's worth keeping an eye out for them to add some variety to the garden.

TAMARISK

(Tamarix)

FEATURES: Deciduous; height to 3 m; spread to 2 m. Non-invasive.

FLOWERS: Feathery, hot pink blossoms all along the branches in spring. A remarkable sight that fairly glows on an overcast day. The flowers are tiny but massed so that the effect is like hot-pink smoke all over the shrub.

LEAVES: Modified to survive very harsh conditions, very small and scale-like. The branches are long and willowy, covered in these downy light green leaves, and have a weeping effect.

WHERE: Anywhere you like! This shrub is resistant to everything — salt spray, wind, drought, insects, heat and cold. Talk about a cast-iron constitution! Oh! and it doesn't like good soil. From the toughest part of the hills to the toughest part of the coast the tamarix will survive.

USES: Absolutely ideal as a windbreak, especially if it has been cut back to encourage bushiness. Fine in a seaside garden, and a very handsome specimen in a regular garden, particularly if it is shaped like a small tree. Prune it as hard as you like, you can't hurt it.

COMMENTS: You would think a shrub as tough as this would be as ugly as a robber's dog, but in fact it has an overall appearance of being soft, and when it flowers it is breathtaking. If you plant one in winter and let it shoot away in spring, you may never have to water it!

TIBOUCHINA

(Tibouchina semidecandra [Lasiandra])

FEATURES: Evergreen; height to 3 m; spread to 2 m. Non-invasive.

FLOWERS: Almost electric, the vivid shades standing out for miles. It seems fitting that such vibrantly coloured flowers should come from such a colourful country as Brazil. This startling array consists of purple, hot pink, soft pink, and a deep rich purple/blue. There is also a white form. The flowers are about the size of a fifty cent coin, have prominent anthers, and cluster together on the terminal growth of branches standing proud. Flowering is over summer.

LEAVES: Elongated oval, thick and fleshy, with an interesting hairy texture, the colour a soft green. In autumn, when the temperatures drop, the odd leaf will turn a rich orange or red. The new growth is often tinged with orange or red.

WHERE: Although they come from the subtropics, they will withstand a mild frost, and so will grow from the hills to the coast. Strong winds can break the brittle branches, but after flowering it is best to cut them back by 50% anyway to stop them becoming too leggy, and this will avoid that problem. A compost-rich soil and a feed of all-purpose fertiliser in spring with good summer watering will keep them happy.

USES: Tibouchina can be trained into a standard over time and will provide a very dramatic effect in the garden if it is given pride of place; likewise it will suit a courtyard or confined area if treated this way. Otherwise grown as an ornamental. *Tibouchina* 'Alstonville', the Australian-bred cultivar, will become a handsome tree if allowed; this is a particularly fine example highly prized around the world.

COMMENTS: Tibouchinas should be grown a lot more than they are; they are so easy to grow and really spectacular when flowering.

CULTIVARS: Cultivar 'Jules' is a compact form with pink to purple flowers and is excellent for a confined area. 'Kathleen' has petals that fold back to reveal a deep pink, and has longer, narrower leaves.

HANDY HINT

LASTING, LEGIBLE LABELS. Nothing could be more frustrating than having a bunch of look-alike plants, such as camellias or azaleas, and not being able to identify the variety because you have lost the label, or it is illegible. Help is at hand. At your local nursery you can purchase an inexpensive pack of aluminium labels thin enough to write on with a ball-point pen. The ink won't stay but the impression of the writing will — forever. Don't tie these onto the plant but place a small stake nearby and tie onto that with the wire thread supplied.

SNAILS AND SLUGS. There are a wide variety of very effective snail and slug baits on the market, unfortunately most of them are as attractive to dogs as they are to pests. If you don't want to spend every weekend at the vet, either buy the special snail pellets which are repellent to dogs or, put your pellets in a jar — without the lid on — in the garden. The snails will crawl in to get the pellets, your dog won't be able to reach them, and when its full, just throw the whole lot out. Of course the easy (and cheap!) alternative is to chuck your snails over the fence into the neighbours' garden.

VIBURNUM

(Viburnum)

FEATURES: Evergreen or deciduous; height to 2 m; spread to 1 m. Non-invasive.

FLOWERS: Mostly white and constructed of heads of many hundreds of tiny flowers that come in various shapes. Guelder Rose (*V. opulus* 'Sterile') is deciduous and produces snowballs of the purest white papery flowers, and *V. opulus* 'Nottcutts' has a lacecap of larger flowers surrounding a mass of tiny central flowers all on a flat head. *V. x burkwoodii* is an evergreen with tiny, waxy white flowers arranged in a perfumed ball. Laurestinus (*V. tinus*) is an evergreen with a grouping of more open fragrant flowers.

LEAVES: The leaves of *V. opulus* are shield-shaped and commonly turn a rich red in autumn, while the evergreens have a tougher, more leathery leaf, oval in shape. The variegated form of *V. tinus* is quite attractive with white/yellow splotches on a green background.

WHERE: Viburnums are easy to grow, requiring a manure-rich soil, deep mulching and good watering to establish. The evergreens, especially *V. tinus*, have proven drought tolerance but will certainly perform better with good watering. The cold doesn't affect them although hot dry winds will burn the leaves of the deciduous varieties. They can be grown from the coast through to the hills, away from violent winds.

USES: Use the deciduous shrubs for accent and display and the evergreens for screening or hedging. As they are amongst some of the most beautiful flowering shrubs in the world, it is worth finding room for one or two.

COMMENTS: *V. opulus* 'Sterile' is a definite favourite of mine. The flowers begin life as a tiny greenish white ball and slowly expand to the size of a snowball, and the whole bush is just covered with them. The perfume of *V. x burkwoodii* is heavenly, worth having one by a back door if you have space. If you are game, have a cool, wind-free part of the garden and live in the hills, try a *V. plicatum* 'Mariesii'. This is a tender, deciduous variety with lacecap flowers floating on top of the branches, and the autumnal colour is stunning.

HANDY HINT

SUBTROPICAL TREES. Subtropical trees can suffer badly in a cold winter. Over summer they make good growth, only to lose it in winter. The experience with poincianas and cassias, in particular, can be heartbreaking. To overcome this problem you can take a number of steps. Firstly, buy the biggest specimen you can find. The idea is get the tree's canopy as high as possible above the ground. Secondly, plant in spring when the ground warms up, as this will give the tree nine months to make a move before the cold sets in. Thirdly, when the tree begins its new growth, give it a good handful of all-purpose fertiliser, around October, and repeat at Christmas. Water the fertiliser in well. Lastly, bear in mind that in their natural habitat they would receive almost daily summer rainfall — so by keeping them very well watered and sprinkling the upper leaves with the hose you will emulate a similar environment.

CREEPERS & CLIMBERS

BUTTERFLY PEA

(Clitoria ternatea)

FEATURES: Evergreen. Non-invasive.

FLOWERS: A beautiful pea flower in the most electric blue, vivid and quite rare in this intensity. The name of the genus comes from the flower shape! C. *ternata* 'Flore Pleno' has double flowers, which only serve to highlight the rich blue. The throat of the flower starts white and turns quickly to a light blue and then to the richest blue on the main petals. Flowering is in summer and autumn.

LEAVES: Small and lance-like in a soft green.

WHERE: This climber enjoys a hot part of the garden and will climb freely to 2 or 3 metres. Coming from the tropics, it doesn't like cold, so plant it where it can get the most sun, facing west or north on a pergola or trellis. Best grown on the coastal plain and near the coast. Clitoria likes rich soil, so give it a feed of all-purpose fertiliser in spring and good watering.

USES: Just as a gorgeous climber and a good talking point at dinner.

COMMENTS: The plant was named by Carl Linnaeus (1707–1778), that master of botanical nomenclature. It would appear he was interested in biology as well.

CLEMATIS

(Clematis)

FEATURES: Deciduous. Non-invasive.

FLOWERS: From the simple, white star-shaped native flower of our bush to the most spectacular mass of purple petals — and a whole range in between. The range of different flower shapes and colours is enormous, and all of them are beautiful. There are good books that illustrate this vast array, and it is worth tracking one down before you choose. The colour range is a bit daunting but you can get white, pink, cerise, red, scarlet, purple, red and white, blue and yellow. Clematis are usually summer flowering.

LEAVES: Usually trifoliate, sometimes with a twining stalk, assisting the climber to scramble upwards.

WHERE: Clematis will grow anywhere away from intense wind. The secret is to keep the roots as cool as possible, ie. plant on the south side of a pergola or trellis, and allow the balance of the climber to work towards the sun that it needs. You can also plant small shrubs around the base to shade it over summer. Plenty of poo and a feed in spring with all-purpose fertiliser, making sure it doesn't dry out.

USES: To highlight a drab area. With such magnificent flowers they would brighten up the dullest spot. The climber can be trained to make sure you have flowers near where you are.

COMMENTS: We had a dead Queensland Box tree, and rather than remove it we planted a clematis by it. The tree is half covered now and in early summer we have thousands of pure white blossoms all over it. Soon we will have a clematis tree! Clematis are very well behaved, can be pruned and I think every home should have one!

CLERODENDRUM

(Clerodendrum)

FEATURES: Evergreen; height 1.5 m; spread 1 m. Non-invasive.

FLOWERS: *C. thomsoniae* is also known as the Bleeding Heart Vine and has very showy white flowers with a prominent red centre, otherwise the clerodendrum are really shrubs. *C. ungandense* has a lovely blue butterfly-shaped flower. The blues are from a powder blue to deep blue in the same flower. *C. inerme* produces white flowers over a very long time, generally from the start of summer into autumn.

LEAVES: Nicely matched to the plant and serve to set the flowers off well — not dense but small and elongated.

WHERE: They will do well anywhere that is frost-free, from the hills to the coast. Give them a warm spot in the garden as they are subtropical and love a bit of heat. Enrich the soil with some poo, and water well to establish.

USES: As an ornamental. They will work well in a tropical-look garden, and *C. thomsoniae* is usually grown in a hanging basket.

COMMENTS: When you look at the flower of *C. ugandense* it makes you marvel at the beauty of nature — such a superbly designed flower.

CYPRESS VINE

(Ipomoea quamoclit)

FEATURES: Evergreen. Non-invasive.

FLOWERS: Very petite, with a red tube about 25 mm long that opens to a perfect five-sided star in a brilliant red. The flowers close at night and then re-open in the morning. Flowering starts in early summer and will continue right through until autumn.

LEAVES: Very delicate and very small, carried on a fine vine. The overall appearance of this creeper is one of gentleness and delicacy. The feather-like leaves are very slender, with lobes about 25 mm long and 1–2 mm across.

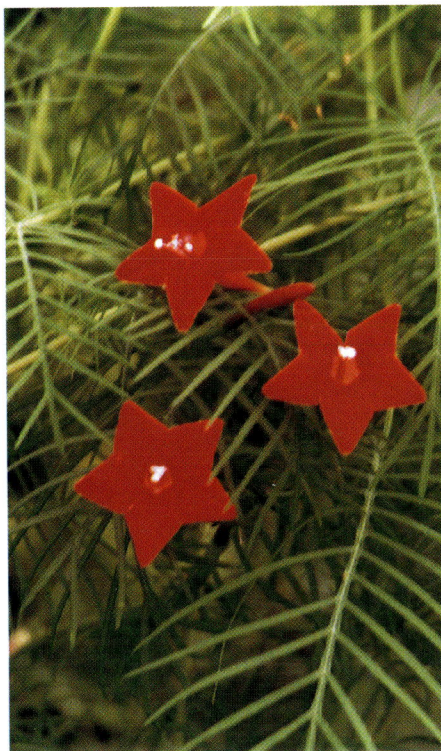

WHERE: Avoid frosts, and grow it in a warm part of the garden. They need something to cling onto, a pergola or some trelliswork. A rich soil with good summer watering will give you a spectacular show of these gorgeous flowers.

USES: This is one of those creepers that won't take over your house and garden like a triffid! It is very well behaved, and probably won't grow beyond a couple of metres or so. It is so fine that it won't cover much, and you could let it climb up an existing vine.

COMMENTS: In really cold weather it may die off, but it will either reshoot, or by then you will have some seeds to plant in early summer. The Mooncreeper or Moonflower (*Ipomoea alba*) is a lovely white-flowered creeper, more vigorous than *I. quamoclit*. They can be planted in tandem, letting them intertwine to achieve a wonderful effect, because one will flower during the day and the other at night!

QUEEN'S WREATH

(Petrea volubilis)

FEATURES: Evergreen. Non-invasive.

FLOWERS: A remarkable collection of mauve/lavender corollas and calyces arranged in long racemes. Flowering over summer, this climber produces such a volume of blossom that it will completely obscure the trellis or pergola on which it is growing. The initial show is from the flowers, but when they are finished the calyces remain to offer a continuation of colour.

LEAVES: Robust and leathery in a darkish green, rounded and paired on the stem.

WHERE: This is a heat-loving climber that needs a west- or north-facing wall or aspect. Best to plant it in spring when the ground is warming up, into a rich soil. Add some well-rotted poo as it is a gross feeder, and fertilise regularly with an all-purpose fertiliser. Don't let it run out of water, and keep the soil moist with a deep mulch of compost. If you live in a frost-prone or very cold area you will need a hot house over winter, so it will need to be grown in a large container and staked. Best grown on the coastal plain, including by the coast itself.

USES: This is one of the most beautiful of all flower displays. Petrea is sometimes called the Purple Wreath, as the mass of flowers is as thick as a wreath, if not thicker. Very useful for one of those hot parts of the garden where nothing else seems to grow, it will need something to twine onto and can be used to hide or obscure a wall or fence.

COMMENTS: The dazzling display put on by this modest climber is breathtaking, something you will never tire of. It can be found from time to time in nurseries, but to get any performance out of it plant it out in spring to give it time to establish before summer takes hold. If your garden is chock-a block, it would be worth removing something to plant a petrea!

SNAIL CREEPER

(*Phaseolus caracalla*) also known as Corkscrew Flower

FEATURES: Evergreen. Non-invasive.

FLOWERS: An interesting form in the shape of a regular garden visitor, the flower is twisted to look for all the world like a large snail shell. They are about the same size as a snail shell, but fortunately have a more pleasing colour. They come in either a light mauve or a mixture of white and mauve. The former has no fragrance, but the latter does.

LEAVES: Arranged in groups of three, elongated, coming to a point, and a soft green.

WHERE: This creeper can be grown from deep shade to full sun and will perform just as well in either situation. It will grow from the coast to the hills and will even take a light frost. It prefers heat and a warm spot in the garden but is really not fussy. The same can be said about the soil; if you build it up the flowering will be better and likewise with watering.

USES: This very hardy creeper won't go crazy in your garden, but if it is happy it will certainly make good growth quickly. This makes it very useful as a screening or hiding creeper. Prune it back in winter if you so desire, as this will encourage secondary growth which will produce more flowers next season.

COMMENTS: The Snail Creeper has such unusual flowers, and over such a long period, that it is well worth finding a spot for it. The perfume is a bonus with the variegated-flowering form. We have one growing up an old gum tree and the effect is just lovely with the flowers cascading down the trunk.

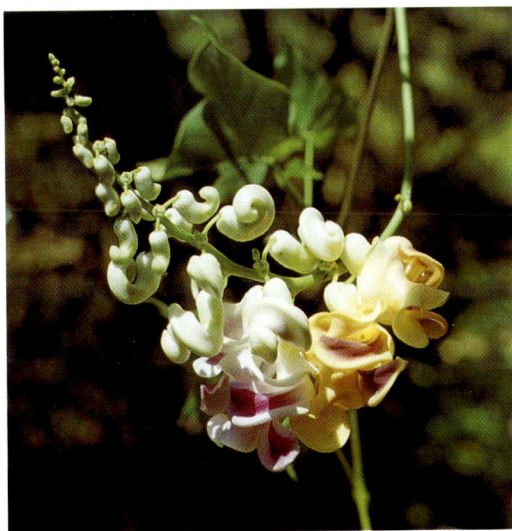

STAR JASMINE

(*Trachelospermum jasminoides*)

FEATURES: Evergreen. Non-invasive.

FLOWERS: Pretty little pure white stars in groups, highly fragrant.

LEAVES: Deep green, rounded, carried in profusion on the vine. The intensity of the leaf colour helps throw the flower into prominence. There is a variegated form that I find particularly attractive. The new growth is white with a tinge of pink, maturing to a leaf that then adds green and they become tricoloured. The effect is superb.

WHERE: A very hardy climber that grows steadily rather than fast, it will take mild frost but definitely prefers a warm position summer and winter. Plant into a good soil and water well to establish. They do show tolerance to long periods without water when older, but it is at the expense of growth. Feed them in spring with an all-purpose fertiliser and again at Christmas. You can grow trachelospermum from the hills to the coast.

USES: This handsome climber can be trained as an espalier and looks superb on a dark brick or limestone wall. Tie wires to the wall either vertically or horizontally, evenly spaced, and then train the climber up it; the wire will be completely covered by the vine and the dark green mass of leaves will be as thick as your arm. It's also very attractive just trained up the post of a pergola and allowed to cascade from the top.

COMMENTS: A prostrate form of this climber can be used very effectively on the top of terraced walls and allowed to waterfall over the edge to the ground. The ease of growing and the depth of field produced by trachelospermum will make it much sought after in the future.

STEPHANOTIS

(Stephanotis floribunda)

FEATURES: Evergreen. Non-invasive.

FLOWERS: Stephanotis have a clustered group of tubular white flowers that are heavily and sweetly perfumed. Usually summer-flowering through to autumn, they will attract the smaller birds to your garden as well as guests.

LEAVES: Thick and fleshy, an elongated oval coming to a point, though sometimes rounded.

WHERE: A heat-loving climber that rambles, though not all that fast growing, it will make steady growth over summer. Best grown on a pergola or trained as an espalier on trelliswork. It is a good idea to thread the new growth through the holes of lattice and keep the lovely blooms spread all over it. You will grow this well anywhere in Perth in a west or north aspect with a rich soil, regular watering and a dose of an all-purpose fertiliser in spring. If you suffer from frost, no worries, grow it in a large pot on a stake and move it against the house in a warm position over winter.

USES: Stephanotis, if trained, will make a good cover over time. Wires woven between two poles of a pergola with the climber trained onto them will hide anything unsightly like diosma.

COMMENTS: The pretty white flowers aside, it is the fragrance that is the real winner. It is a little like jasmine, but in a class of its own.

PHOTOGRAPHS

18 *Albizia julibrissin*
19 *Bauhinia alba*
20 *Acer negundo*
21 *Magnolia grandiflora*
22 *Pistacia chinensis*
23 *Sapium sebiferum*
25 *Fraxinus oxycarpa* 'Raywood'
26 *Dombeya macrantha*
27 *Ficus benjamina*
28 *Prunus persica* 'Albopiena'
29 *Prunus x blireiana*
30 *Gleditsia triacanthos* 'Sunburst'
31 *Laburnum x watereri* 'Vossii'
33 *Ulmus procera* 'Louis van Houtte'
34 *Schizolobium parahybum*
35 *Koelreuteria paniculata*
36 *Cassia fistula*
37 *Crataegus monogyna*
38 *Brachychiton acerifolius*
39 *Jacaranda mimosaefolia*
40 *Acer palmatum*
41 *Cercis siliquastrum*
42 *Harpephyllum caffrum*
43 *Salix caprea*
44 *Lagerstroemia indica*
45 *Liquidambar styraciflua*
46 *Platanus x acerifolia*
47 *Magnolia x soulangiana*
49 *Gingko biloba*
50 *Pyrus ussuriensis*
51 *Hymenosporum flavum*
52 *Azadirachta indica*
53 *Celtis sinensis*
54 *Parrotia persica*
55 *Quercus palustris*
56 *Robinia pseudoacacia* 'Decaisneana'
57 *Paulownia kawakami*
58 *Malus ioensis* 'Plena'
59 *Delonix regia*
60 *Chorisia speciosa*
61 *Betula pendula*
62 *Populus nigra* 'Italica'

63 *Cotinus coggygria*
64 *Spathodea campanulata*
65 *Taxodium distichum*
66 *Tabebuia rosea*
67 *Tipuana tipu*
68 *Toona australis*
69 *Nyssa sylvatica*
70 *Pittosporum tenuifolium* 'Irene Paterson'
71 *Prunus subhirtella* 'Pendula Rosea'
72 *Morus alba* 'Pendula'
73 *Sophora japonica* 'Pendula'
76 *Alberta magna*
77 *Alternanthera dentata*
78 *Berberis thunbergii* 'Rose Glow'
79 *Buxus sempervirens*
80 *Cytisus scoparius* 'Andreanus'
81 *Brunfelsia latifolia*
82 *Buddleja salvifolia*
83 *Orthosiphon aristatus*
85 *Camellia sasanqua* 'Chansonette'
86 *Ceanothus impressus* 'Blue Pacific'
87 *Cestrum* 'Newellii'
88 *Vitex agnus-castus*
89 *Choisya ternata*
90 *Convolvulus cneorum*
91 *Cotoneaster microphylla*
92 *Daphne alpina*
93 *Daubentonia tripetti*
94 *Deutzia scabra* 'Candissima'
95 *Dichorisandra thyrsiflora*
96 *Duranta repens*
97 *Punica granatum* 'Flore Pleno'
98 *Echium candicans*
99 *Escallonia bifida*
100 *Euphorbia leucocephala*
101 *Chaenomeles speciosa* 'Moerloosii'
102 *Gardenia augusta* 'Golden Magic'
103 *Garrya elliptica*
104 *Forsythia suspensa*

105 *Gordonia axillaris*
107 *Hibiscus syriacus*
108 *Ilex aquifolium* 'Golden Milkboy'
109 *Lonicera japonica*
111 *Hydrangea macrophylla*
112 *Inga edulis*
113 *Rhaphiolepis x delacourii*
114 *Indigofera australis*
115 *Cephalotaxus harringtonia*
116 *Lavandula canariensis*
117 *Syringa vulgaris* 'Sensation'
118 *Clethra arborea*
119 *Mahonia lomariifolia*
120 *Lavatera acerifolia*
121 *Romneya coulteri*
123 *Michelia doltsopa*
124 *Prostanthera ovalifolia*
125 *Philadelphus x virginalis*
126 *Myrtus communis*
127 *Tetradenia riparia*
128 *Ochna serrulata*
129 *Murraya paniculata*
130 *Osmanthus fragrans*
131 *Osmoxylon lineare*
133 *Photinia glabra*
134 *Phyllanthus acidus*
135 *Euphorbia pulcherrima*
137 *Rhododendron* 'Sappho'
138 *Cistus x canescens* 'Albus'
139 *Breynia nivosa*
140 *Euonymus fortunei* 'Silver Pillar'
141 *Tamarix aphylla*
143 *Tibouchina* 'Alstonville'
145 *Viburnum opulus* 'Sterile'
148 *Clitoria ternatea*
149 *Clematis* 'Belle of Woking'
150 *Clerodendrum ungandense*
151 *Ipomoea quamoclit*
152 *Petrea volubilis*
153 *Phaseolus caracalla*
154 *Trachelospermum jasminoides*
155 *Stephanotis floribunda*

REFERENCES

Coombes, A. (1992). *World Trees: Collins Eyewitness Guides.* Harper-Collins.

Courtright, G. (1988). *Tropicals.* Timber Press.

Ellison, D. (1996). *Cultivated Plants of the World.* Flora Publications.

Huxley, A. (ed.) (1992). *The New Royal Horticultural Society Dictionary of Gardening.* 4 vols. Macmillan.

Macaboy, S. (1986). *What Flower is That?* Weldon Publishing.

Macaboy, S. (1989). *What Shrub is That?* Lansdowne.

Old, M. (ed.) (1994). *The Ultimate Australian Gardening Book.* Random House.

INDEX

Acer
  negundo 20
  negundo 'Golden Variegated Box' 20
  negundo 'Kelly's Gold' 20
  negundo 'Silver Variegated Box' 20
  palmatum 40
Alberta magna 76
Albizia julibrissin 18
Albizzia 18
Alternanthera 77
Alternanthera
  dentata 77
  dentata 'Apricot Splash' 77
Apple-blossom Cassia 36
Apricot, Weeping, 71
Ash
  Claret 24-5, 107
  Evergreen 24
  Flowering 24
  Variegated 24
  Weeping Golden 24
Aspen, Trembling 62
Azalea 136
Azadirachta indica 52

Barberry 78
Bauhinia 19
  White 19
Bauhinia
  alba 19
  galpinni 19
  variegata 'Candida' 19
Bells, Golden 104
Berberis
  atropurpureum 78
  atropurpureum 'Golden Ring' 78
  atropurpureum 'Rose Glow' 78
Betula
  pendula 61
  pendula 'Dalecarlica' 61
  pendula 'Youngii' 61
Birch
  Cut-leaf 61
  Silver 61
  Weeping 61
Bleeding Heart Vine 150
Bo Tree 27
Bolivia, Pride of 67
Box 79
Box Elder Maple 20
Brachychiton

acerifolia 38
populneus 38
Breynia
  nivosa 139
  nivosa 'Iron Range' 139
  nivosa 'Rosea-Picta' 139
Broom 80
Brunfelsia 81
Brunfelsia latifolia 81
Buddleia, 31
Buddleja
  davidii 82
  latifolia 82
bulbs 51, 67, 90
Bull Bay Magnolia 21
Butterfly Bush 82
Butterfly Pea 148
Buxus
  sempervirens 79
  latifolia 79

Camellia 84-85
Camellia
  chrysantha 84
  'Mandy' 84
  japonica 84
  sasanqua 84
Cape Wedding Bush 26
Cassia 33, 36
  Apple-blossom 36
Cassia
  fistula 36
  javanica 36
  nodosa 36
Cat's Whiskers 83
Ceanothus
  impressus 'Blue Cushion' 86
  impressus 'Blue Pacific' 86
Cedar, Chinese 68
Cedrella sinensis 68
Celtis australis 53
Cephalotaxus harringtonia 115
Cercis
  canadensis 41
  siliquastrum 41
Cestrum 87
Cestrum
  newelli 87
  nocturnum 87
  purpureum 87
Chaenomeles speciosa 101
Chain, Golden 31
Chaste Tree 88
Chinese Cedar 68

Chinese Elm 32
Chinese Pistachio 22
Chinese Tallow 23
Choisya
  ternata 89
  ternata 'Sundance' 89
Chorisia speciosa 60
Cistus 138
Claret Ash 24-5, 107
clay soil 81
Clematis 149
Clerodendrum
  inerme 150
  thomsoniae 150
  ungandense 150
Clethra arborea 118
climate 11, 145
Climbing Hydrangea 110
Clitoria ternatea 148
  'Flore Pleno' 148
Contorted Willow 43
Convolvulus cneorum 90
Cork Oak 55
Corkscrew Flower 153
Cotinus
  coggygria 63
  coggygria 'Purpureus' 63
  coggygria 'Royal Flame' 63
Cotoneaster, 91
Cotoneaster
  dammeri 91
  horizontalis 'Variegatus' 91
  microphylla 91
Cottonwood 62
Crab-Apple, Prairie 58
Crataegus
  laevigata 'Paul's Scarlet' 37
  monogyna 37
Crepe Myrtle 44
Cricket Bat Willow 43
Cut-leaf Birch 61
Cypress Oak 55
Cypress Vine 151
Cypress, Swamp 65
Cytisus 80

Daphne 92
Daphne odora 92
Daubentonia tripeti 93
Delonix regia 33, 34, 59
Deutzia
  gracilis 94
  scabra 'Candissima' 94
Dichorisandra thyrsiflora 95

Dombeya 26
Dombeya
  macrantha 26
  natalensis 26
Duranta
  erecta 96
  erecta 'Aussie Gold' 96
  erecta 'Sheena's Gold' 96
  repens 96
Dwarf Flowering Pomegranate 97
Dutch Elm Disease 69

Echium
  candicans 98
  fastuosum 98
Elm
  Fastigiate 32
  Silver 32
  Weeping 32
  Chinese 32
  Golden 32-33
  Upright 32
English Box 79
English Oak 55
Escallonia bifida 99
Euonymus
  fortunei 140
  fortunei 'Emerald 'n' Gold' 140
  fortunei 'Emerald Gaiety' 140
  fortunei 'Silver Pillar' 140
Euphorbia
  leucocephala 100
  leucocephala 'Pink Finale' 100
  pulcherrima 135
  pulcherrima 'Henrietta Ecke' 135
Evergreen Ash 24

Fastigiate Elm 32
fertilising 14
Ficus 27
Ficus
  benjamina 27
  macrocarpa 27
  religiosa 27
Fig, Moreton Bay 27
Flame Tree, Illawarra 38
flies 42
Flowering Ash 24
Flowering Peach 28
Flowering Plum 29
Flowering Quince 101
Fly-trap, Venus 25
Forsythia
  x *intermedia* 104
  suspensa 104
Frangipani, Native 51

Fraxinus
  'Golden Ash' 24
  griffithii 24
  oxycarpa 'Raywood' 24-25
  'Variegated Ash' 24
  'Weeping Golden Ash' 24

garden as an asset 8
Gardenia 102
Gardenia
  augusta 'Florida' 102
  augusta 'Golden Magic' 102
  augusta 'Professor Pucci' 102
  augusta 'Radicans' 102
  'Ocean Pearl' 102
  thunbergia 102
Garrya elliptica 'James Roof' 103
Gingko biloba 48
Gleditsia
  triacanthos 'Rubylace' 30
  triacanthos 'Shademaster' 30
  triacanthos 'Sunburst' 30
Gleditsia, Sunburst 30
Golden Ash 24
Golden Bells 104
Golden Elm 32-33
Golden Oak 55
Golden Poinciana 33, 34
Golden Rain 35
Golden Shower 36
Golden Chain 31
Golden Variegated Box Maple 20
Gordonia axillaris 105
Guelder Rose 144

hardiness 8
Harpephyllum caffrum 42
Hawthorn 37
  Indian 113
Heaven Scent Magnolia 47
Hibiscus 106-107
Hibiscus
  mutabilis 106
  rosa-sinensis 106
  schizopetalus 106
  syriacus 106
Holly 108
  Weeping 108
Honeysuckle 109
Hong Kong Orchid 19
Hortensia 110
humus 11
Hydrangea 110-111
  Climbing 110
  Oak Leafed 110
Hydrangea
  petiolaris 110

quercifolia 110
Hymenosporum flavum 51

Iboza riparia 127
Ice-cream Bean Tree 112
Ilex
  altaclerensis 'Golden King' 108
  aquifolium 'Argentea Marginata' 108
  aquifolium 'Golden Milkboy' 108
  aquifolium 'Lawsonia' 108
  cornuta 108
Illawarra Flame Tree 38
Indian Hawthorn 113
Indigo Bush 114
Indigofera decora 114
Inga edulis 112
Ipomoea
  alba 151
  quamoclit 151

Jacaranda 39
Jacaranda
  mimosaefolia 39
  mimosaefolia 'Variegated' 39
  mimosaefolia 'White' 39
Japanese Maple 40
Japanese Plum-Yew 115
Japonica 101
Jasmine, Star 154
Jessamine, Orange 129
Judas Tree 41

Kaffir Plum 42
Kilmarnock Willow 43
Koelreuteria paniculata 35
Kurrajong 38

labels 133, 143
Laburnum x *watereri* 'Vossii' 31
Lagerstroemia 44
Lagerstroemia indica 44
Lasiandra 142
Laurestinus 144
Lavandula
  canariensis 116
  canariensis 'Sidonie' 116
Lavatera acerifolia 120
Lavender 116, 126
Leopard Tree 33
light 76
Ligustrum 117

Lilac 117
Lily-of-the-valley 118
Liquidambar 45
  Variegated 45
Liquidambar
  styraciflua 45
  styraciflua 'Canberra Gem' 45
  styraciflua 'Jennifer Carroll' 45
  styraciflua 'Palo Alto' 45
London Plane 46
Lonicera
  etrusca 'Superba' 109
  hildebrandiana 109
  japonica 109
  nitida 109

Magnolia 47
  Bull Bay 21
  Portwine 122
Magnolia
  grandiflora 'Exmouth' 21
  x *soulangiana* 47
  x *soulangiana* 'Heaven Scent' 47
  stellata 47, 122
Mahonia 119
Mahonia
  aquifolium 119
  lomariifolia 119
Maidenhair Tree 48
Mallow 120
Malus
  'Echtermeyer' 58
  'Golden Hornet' 58
  ioensis 'Plena' 58
  'Purpurea' 58
Manchurian Pear 50
manure (poo) 11-12
Maple
  Box Elder 20
  Golden Variegated Box 20
  Japanese 40
  Kelly's Gold 20
  Silver Variegated Box 20
Matilija Poppy 121
Michelia 122-123
Michelia
  champaca 122
  doltsopa 122
  figo 122
Mint Bush 124
Mock Orange 125
Mooncreeper 151
Moonflower 151
Mop-top Robinia 56
Moreton Bay Fig 27
Morus alba 'Pendula' 72
Mulberry, Weeping 72
mulch 11-12

Murraya
  alata 129
  paniculata 129
Myrtle 126
Myrtle , Crepe 44
Myrtus
  communis 126
  luma 126

Native Frangipani 51
Neem Tree 52
Nettle Tree 53
New Zealand Hybrid Willow 29
Nursery Industry Association of
  Western Australia 8-9, 15
Nutmeg Bush 127
Nyssa sylvatica 69

Oak
  Cork 55
  Cypress 55
  English 55
  Golden 55
  Pin 55
  Willow 55
Oak Leafed Hydrangea 110
Ochna serrulata 128
Orange Jessamine 129
Orange, Mock 125
Orchid, Hong Kong 19
Oriental Plane 46
Orthosiphon aristatus 83
Osmanthus
  fragrans 130
  heterophyllus 'Variegatus' 130
Osmoxylon lineare 130

Parrotia persica 54
Passionfruit 54
Paulownia
  fortuneii 57
  kawakami 57
Pea, Butterfly 148
Peach
  Flowering 28
  Weeping 28
Pear
  Manchurian 50
  Silver 50
  Willow-leaf 50
Pepperidge 69
Persian Silk Tree 18
Persian Witch-Hazel 54
Petrea volubilis 152
Phaseolus caracalla 153

Philadelphus
  coronarius 125
  x *lemoinei* 125
  x *virginalis* 125
Photinia 132-133
Photinia
  beauvardiana 132
  x *fraseri* 132
  x *fraseri* 'Robusta' 132
  x *fraseri* 'Rubens' 132
  glabra 132
  'Superhedge' 132
Phyllanthus minutiflorus 134
Pieris japonica 118
Pin Oak 55
Pink Wisteria Tree 56
Pistacia chinensis 22
Pistachio, Chinese 22
Pittosporum 70
Pittosporum
  eugenioides 'Variegatum' 70
  tenuifolium 'Irene Paterson' 70
  tenuifolium 'Limelight' 70
  tenuifolium 'Tom Thumb' 70
Plane
  London 46
  Oriental 46
Platanus
  orientalis 46
  x *acerifolia* 46
Plum
  Flowering 29
  Kaffir 42
  Red-leaf 29
Plum-Yew, Japanese 115
Poinciana
  Golden 33, 34
  Royal 33, 34, 59
Poinsettia 135
Pomegranate, Dwarf Flowering 97
Poplar, Simon's 62
Poppy, Matilija 121
Populus
  deltoides 62
  simonii 62
  tremuloides 62
Portwine Magnolia 122
pots, terracotta, 53, 71
potting mixes 14-15
Powton 57
Powton Sapphire Dragon 57
Prairie Crab-Apple 58
Pride of Bolivia 67
Prostanthera
  cuneata 'Alba' 124
  ovalifolia 124
  ovalifolia 'Alba' 124
  rotundifolia 124
Prunus
  cerasifera 'Nigra' 29
  mume 'Pendula' 71

persica 'Albopena' 28
persica 'Pendula' 28
persica 'Versicolor' 28
wrightii 29
x *blireiana* 29
Punica granatum nana 97
Purple Wreath 152
Pyrus
  salicifolia 'Pendula' 50
  'Totem Pole' 50
  ussuriensis 50
  'Winterglow' 50

quarantine 69
Queen's Wreath 152
Quercus
  palustris 55
  phellos 55
  robur 55
  robur 'Concordia' 55
  robur 'Fastigiata' 55
  suber 55
Quince, Flowering 101

Rain, Golden 35
Redbud 41
Red-leaf Plum 29
Rhaphiolepis
  indica 113
  indica 'Springtime Spring
Song' 113
  umbellata 113
  x *delacourii* 113
Rhododendron 136-137
Rhus Tree 22
Robinia
  pseudoacacia 'Decaisneana' 56
  pseudoacacia 'Frisia' 56
  pseudoacacia 'Mop-top' 56
  uniflora 56
Rock Rose 138
Romneya coulteri 121
Rose of Sharon 107
Rose, Rock 138
Royal Poinciana 33, 34, 59

Salix
  alba caerulea 43
  caprea 'Kilmarnock' 43
  matsudana 'Tortuosa' 43
Sapium sebiferum 23
Scarlet Wisteria Tree 93
Schizolobium parahybum 34
secateurs 138
seeds, importing 69
Sesbania grandiflora 93

Sharon, Rose of 106
Shower, Golden 36
Silk Tree, Persian 18
Silk-floss Tree 60
Silver Birch 61
Silver Elm 32
Silver Pear 50
Silver Trumpet Tree 66
Silver Variegated Box Maple 20
Simon's Poplar 62
Smoke Tree 63
Smokebush 63
Snail Creeper 153
snails 143
Snowbush 139
Snowflake 100
soil 10-15, 81, 99
Sophora japonica 'Pendula' 73
Sophora, Weeping 73
South African Tulip Tree 64
Spathodea campanulata 64
Spindle-Tree 140
staking 85, 123
Star Jasmine 154
Stephanotis floribunda 155
Sunburst, Gleditsia 30
Swamp Cypress 65
Sweet pea 123
Syringa 117

Tabebuia 66
Tabebuia
  argentea 66
  rosea 66
Tallow, Chinese 23
Tamarisk 141
Tamarix 141
Taxodium distichum 65
terracotta pots 53, 171
Tetradenia riparia 127
Tibouchina 142-43
Tibouchina
  semidecandra 142
  semidecandra 'Alstonville' 142
  semidecandra 'Jules' 142
  semidecandra 'Kathleen' 142
Tipuana 67
Tipuana tipu 67
Toona 68
Toona australis 68
Tortured Willow 43
Trachelospermum jasminoides 154
Trembling Aspen 62
Trumpet Tree, Silver 66
Tulip Tree, South African 64
Tupelo 69
Ulmus
  parvifolia 32
  procera 'Louis van Houtte' 32

procera 'Silver' 32
procera 'Upright' 32
procera 'Weeping' 32
Upright Elm 32

Variegated Ash 24
Variegated Jacaranda 39
Variegated Liquidambar 45
Variegated Pittosporum 70
Venus Fly-trap 25
Viburnum 144-45
Viburnum
  opulus 'Nottcutts' 144
  opulus 'Sterile' 144
  plicatum 'Mariesii' 144
  tinus 144
  x *burkwoodii* 144
Vine, Cypress 151
Vitex agnus-castus 88

watering 13-14
Wedding Bush, Cape 26
Weeping Apricot 71
Weeping Birch 61
Weeping Elm 32
Weeping Golden Ash 24
Weeping Holly 108
Weeping Mulberry 72
Weeping Peach 28
Weeping Sophora 73
Weeping White Wisteria Tree 56
White Bauhinia 19
White Jacaranda 39
White Wisteria Tree, Weeping 56
Willow
  Contorted 43
  Cricket Bat 43
  Kilmarnock 43
  New Zealand Hybrid 43
  Tortured 43
Willow Oak 55
Willow-leaf Pear 50
Wisteria Tree
  Pink 56
  Scarlet 93
  Weeping White 56
Witch-Hazel, Persian 54
worm farms 118
Wreath, Queen's 152

Yesterday, Today & Tomorrow 81
Yew
  Irish 115
  Japanese Plum 115